Ting Wen-chiang

HARVARD EAST ASIAN SERIES 42

The East Asian Research Center at Harvard University
administers research projects designed to further
scholarly understanding of China, Japan,
Korea, Vietnam, and adjacent areas.

Ting Wen-chiang

Science and China's New Culture

CHARLOTTE FURTH

HARVARD UNIVERSITY PRESS

Cambridge, Massachusetts

1970

Preparation of this volume has been aided by
a grant from the Ford Foundation.

Library of Congress Catalog Card Number 78-95920
SBN 674-89270-4

Printed in the United States of America

Author's Note

Ting Wen-chiang, the man this book is about, is difficult to categorize. A geologist by profession, he was perhaps the best known of the few pioneer Western-trained scientists in China during the tumultuous years of the Chinese republic. But he was also an official, journalist, businessman, polemicist, and educator — a leader of the Peking academic establishment in the 1920s and 1930s and one of the important personalities of China's "new culture" movement. To all of his varied activities he brought the perspective of a wide-ranging scientific modernizer during the first decades of the twentieth century, when wholesale "Westernization" of China was the ambition of a broad, reformist intelligentsia. In addition, he was a man of Confucian training and gentry background who tried to make the career of a modern scientist function according to some of the patterns prescribed by that tradition. In sum, two endeavors dominated Ting's career and gave it unity: first, the struggle to understand modern science and its impact upon the old Chinese intellectual order; and second, the search for methods of meaningful action on the part of an elite still haunted by the imperatives of Confucian scholar statesmanship. In writing about him, these are the themes I have tried to emphasize.

The first four chapters of the book deal largely with Ting's purely scientific career; his political beliefs and actions are the subject of the last three. Such an arrangement brings out most clearly the problems of Ting's intellectual development, but at the cost of taking a few liberties with his chronological story. To help the reader, I have provided in Appendix A a year-by-year list of the chief events in Ting's own life and the public happenings which most influenced it.

I am grateful to Professor James T. C. Liu, who first suggested Ting to me as a subject. He, together with Professors David Nivison and Lyman Page Van Slyke, originally guided the manuscript as it was written at Stanford University, and their help was fundamental to its creation. Among the many other members of the profession who have offered suggestions and aid, I am particularly indebted to Irene Eber, Phillip Huang, Benjamin Schwartz, and John K. Fairbank, and to Chia Fu-ming and Yang Chün-shih. Harvard's East Asian Research Center provided a welcome haven during the months when the manuscript was being prepared for press, and the Social Science Research Council provided a grant without which I could not have written. My special thanks goes to Mrs. Olive Holmes for her care and patience in helping prepare the bibliography. The staffs of the Hoover Institute, the East Asiatic Library at Berkeley, the Oriental Library at U.C.L.A., the Harvard-Yenching Library, and the Library of Congress, who gave me ungrudging cooperation and time, deserve my thanks also. By here expressing gratitude

AUTHOR'S NOTE

to my husband, Montgomery Furth, for his personal support and critical suggestions, I am making a conventional gesture in a situation where conventions cannot measure debt.

Charlotte Furth

May 1969

Contents

Contents

Ting Wen-chiang

Introduction

China was called a republic for thirty-seven years, from 1912, when the Manchu dynasty finally abdicated, to the victorious Communist revolution of 1949. Since thirty-seven years is little more than half a lifetime, individuals whose active careers were concentrated in that interval were born before the republic came into existence, and many survived long after it was over. The abrupt change and concentrated pace made Republican reformers self-conscious, often frustrated experimentalists, attempting to mold a society still deeply conditioned to the habits of imperial Confucianism. Although Western models dominated reformist thinking, their application met with many obstacles, both inherited from past tradition and imposed by the anarchic conditions of Republican politics.

Yet the failure of the republic as an institutional experiment, however tragic, was stimulating to the intellectual life of the time, creating an atmosphere of dynamic speculation and debate among men freed by social chaos from the restraints as well as the comforts of orthodoxy. Because of the historical association of scholarship and power in Confucian China, the intelligentsia who led the way in these debates formed a particularly sensitive group within the Republican leadership. As an educated elite, they stood as heirs to the scholar-bureaucrats of the old empire, and, at the same time, their function as scholars made them pioneers in the study of Western ideas and catalysts of innovation.

The position of such men, standing between two worlds, is illustrated on the most basic level by the pattern of their education. Under the republic an intellectual of any importance was not merely a student of Western culture but a graduate of a foreign university. Nevertheless, this generation was also the last certain to have been exposed to the Chinese classics and, more important, to the entire Confucian view of the function of knowledge in society. That view linked the realms of thought and action, teaching men to honor learning as fundamental and to associate knowledge with the right to moral and political leadership. Through the imperial examination system, scholarship led directly to political action, and when politics was presumed to be in the hands of an ethically indoctrinated cultural leadership, educated men felt little sense that there need be any basic antagonism between their own highest values and state power. Thus, although the Chinese graduate of a Western university in the early twentieth century was equipped with the intellectual tools provided by a Western scientific or social-scientific curriculum, he often continued to view the career of an educated man in this Confucian perspective. In return, the Chinese public saw him the same way, as enjoying the status and responsibilities of a successful imperial degree holder.

The fact that advanced education in Western subjects was still acquired abroad had important social consequences. It guaranteed that foreign university degrees were available only to the children of the privileged minority which could afford the expense and that they would be earned in specialized subjects like law, engineering, or medicine, for which China did not yet have an appropriate professional context. Back in China, foreign graduates found it hard to function either as a traditional elite, in the manner suggested by their background, or as effective professional men in the fields of their training. Drawn from families with historical pretensions to social leadership, exposed to the traditional high culture and its values, they nevertheless enjoyed more sophisticated and first-hand experience of Western countries than any Chinese ever had before. It made them more aware of the complexity of the West itself and of the full implications of China's material backwardness, more critical of the Chinese past, and, thus, torn between angry iconoclasm and nostalgic attachment to a great cultural heritage. At the same time it left them oddly isolated from many of the immediate problems of China's peasant masses, not only because of the perennial gulf between the educated and the illiterate, but also because of their novel cosmopolitanism, their urban way of life, and their political weakness. In contrast with earlier scholar generations, Republican intellectuals were becoming a relatively alienated group, concentrated in a few universities, government bureaus, and commercial seaports.[1]

China's "new culture movement" was created by such men. Its ideas were first discussed among Chinese students in Japan in the last years of the dynasty, and it reached its widest audience among the educated vanguard during the republic's first decade. As students themselves and men poised between contrasting civilizations, the movement's leaders saw the irony in their effort to create a "culture" overnight, to develop an entire outlook on the world by conscious selection rather than gradual habituation. Since they were intelligentsia, they found it easiest to be effective where the enterprise was itself academic. The great immediate achievement of the new culture movement was the reform of the literary language, which by 1920 or so made it possible for people to communicate on paper in the words of everyday life, and which laid the foundations for eventual mass literacy. The movement's most sustained practical effort was the prolonged examination, in and out of the academies, of all aspects of Western culture, and the establishment of a new Chinese scholarship, critical of native classical canons and in harmony with contemporary worldwide research.

However, the new culture movement involved more than academic reorientation; it also stood for an idealized, programmatic vision of a new Chinese man and society, and as such it fertilized the increasingly radical political action movements of the day.

The theorists of the new culture assumed that societies develop organically as the expression of certain fundamental "principles" — moral, social, and natural. Thus Confucian society had always been described, and so the new culture was expected to flower when key Western principles had been assimilated by Chinese, whose creative forces would thereby be released. The new culture, like the old, had to begin with a kind of man. The movement exalted Western liberalism, nationalism, democracy, and science, each of which was seen in terms of human attitudes generating them, attitudes now needed to effect the moral rearmament of the Chinese consciousness. Chinese were to become personally emancipated, freed from the restraints of the Confucian family, the subjection of filial piety and arranged marriages, the indignities of footbinding and concubinage — people capable of creative work and with a hope for personal happiness. They were to become filled with a new spirit of modern patriotism, which would mobilize them as citizens to free China from the humiliations of foreign exploitation; they were to become active participants in public affairs so that the nation might enjoy unity and freedom from civil war; and lastly, they were to become filled with the spirit of modern science, which would guarantee the rationality of their thoughts and the economic and technological productivity of their labor.

Technically speaking, the new culture movement was the outburst of only a few years, between approximately 1915 and 1923. Then its classic manifestos were written, and the words *"hsin wen hua"* served as a banner for the entire radical intelligentsia. But the assumptions of the movement had had a longer history: the belief that fundamental cultural change was a prerequisite for successful modernization and that the models should come from the West went back to 1902 and Liang Ch'i-ch'ao's call for a "new people." [2] Moreover, problems raised by these assumptions outlasted the high tide of the movement as well as contributing to schisms within it. What should be one's standard for selection? This and like questions divided all-out "Westernizers" from East-West synthesizers and split pragmatists from ideologues. How could the new culture be made a reality? Here evolutionists who relied on the power of a new education to fertilize the natural growth of modern institutions differed from activists impatient to follow theoretical blueprints and even to wage revolutionary war for the necessary political power. In this way the issues of the new culture movement reverberated through the 1920s and 1930s, contributing to the awakening political consciousness of the Chinese people and to the rise of Communism. Its assumptions fed that persistent strand of modern Chinese thought which has placed the restructuring of the Chinese consciousness, down to the root of moral and social attitudes, at the heart of China's modern revolution.

5

As a man of this Republican generation of Western-educated intelligentsia, Ting Wen-chiang was deeply involved in the new culture movement. For much of his adult life he was busy fighting its battles, and in his person he exemplified for many Chinese certain of its dominant ideals. Born in rural Kiangsu of gentry parents, he was educated in England as a geologist and came back to China in 1911, only two months before the outbreak of the Wuchang revolution. In the world of foreign graduates under the early republic, he was a versatile and commanding figure, founder and first director of China's Geological Survey, then in the early 1920s a private businessman, manager of a coal mine in Jehol. He entered politics briefly as the director-general of the city of Greater Shanghai under a warlord government in 1926, and in later life returned to science as an academic, becoming professor of geology at Peking University and Secretary-General of the Academia Sinica, Nationalist China's principal scholarly research institute. Throughout the 1920s and 1930s he was a writer and editorial adviser for various intellectual and political journals, particularly *Endeavor* (*Nu-li chou-pao*) and the *Independent Critic* (*Tu-li p'ing-lun*), and a close, influential friend of such spokesmen for the new culture movement as Hu Shih, Liang Ch'i-ch'ao, and Fu Ssu-nien.

However, Ting's special place in the history of the new culture movement is as its most prominent scientist. He was one of the first Chinese who had studied science in the West from both technical and philosophical points of view, and a man who felt a personal responsibility to educate his countrymen in the principles of scientific thought. In his experience can be seen both the attraction which "Mr. Science" had for radical Chinese and the problems of making scientific practice possible or scientific concepts intelligible in a Chinese setting. The role Ting aspired to — scientist as cultural and political leader — was entirely novel in Chinese historical experience.

Little in Chinese tradition offered any preparation for the ideas of Western science. It is commonly remarked that traditional Confucianism was "humanistic": devoted to the study of "men and affairs," and centered around a primarily ethical and socially oriented classical canon. This dominant scholarly tradition left educated Chinese little scope for the observation of nature and even less incentive for practical invention — activities presumed to have contributed substantially to the growth of science in the West. Perhaps even more important for the nondevelopment of science in China was the idea of nature which unconsciously molded ordinary Chinese conceptions of the cosmos — a cosmological view which did not spring directly from the original Confucian writings, but which was compatible with them and actually became integrated into the orthodox imperial and neo-Confucian philosophical traditions. Modern scholars like Joseph Needham have dubbed this view "organic," and opposed it to the "mechanistic" cosmologies of the West.[3] It proceeded from the concept of the Tao, the potent negative source of the entire multiplicity of things — a generating force held responsible for the existence of the universe, and discernible only at work — in the natural relations of things in a patterned, functioning whole. It is dangerous to push a biological analogy too far: suffice it to say that this view of nature emphasized unseen forces behind the phenomena of the world without assuming that those forces were transcendent or in any way basically separable from the activities of nature itself, and that its model for the motions of the universe was the model of the yin and yang — harmonious alternating activity in which, as in an organism, the function cannot be separated from the structure. In Western terms it is clear that such a view is not a view of nature only. Without the concept of the supernatural nothing is merely natural in the Western sense of "pertaining to the physical world." In the West "physics" has been opposed to "metaphysics," and this contrast is reflected in the historically important philosophical dualism between matter and spirit. In Chinese thought, people were not disposed to set apart the world of nature, separate from an essentially spiritual man and from a transcendent God, and to see it as an object — to be created by deity and contemplated and ultimately manipulated by man. They saw man the social being as an integral part of a cosmic order whose fundamental "principles" were at the same time both natural and normative. Such "principles" were like the nourishing root of all things, generating the overarching pattern of heaven, but also embedded in man's moral nature, and by their power creating through man the arrangements of human society. To neo-Confucian philosophers the moral purposes of the universe were linked with its natural activities: man loved virtue in accordance with the same Way that set the sun in its zodiacal course. Man and the universe were bound in a natural, yet more than naturalistic, harmony.

Now science in the West is derived from more than a view of nature as unmagically material and mechanistic, the potential object of observation. It is above all the outcome of a methodology not yet completely understood, but rooted in European disciplines of logic and mathematics. Since Greek times, formal, logically cohesive systems of reasoning, constructed from the abstract concepts of logic and mathematics, have been regarded in Europe as a rational ideal. The formative age of scientific discovery began with Newtonian mechanics, when this Euclidean model of a perfectly coherent logical system was linked with the notion of experimental verifiability. Once sophisticated mathematical language was used to express relations interpretable in terms of empirical observation, scientific theories came into being, that is, theories both logically cohesive and capable of elucidating facts about the real world.

Traditional Chinese, however, when considering the methodology of sound reasoning, had no parallel intellectual heritage. In matters beyond the verification of the senses the question of whether to accept or reject a proposition was not usually thought of as a problem solvable by the application of logical categories. This is not to say that Chinese thinkers did not reason. Their preferred methods for making and justifying assertions followed entirely plausible paths. First and foremost they appealed to history: what had been asserted in the past had a privileged claim to respect. Original thinkers often justified their innovations by manipulating and restructuring accepted concepts of the past. Second they used the method of syncretism: people preferred to see the holistic aspect of things, to distinguish concepts, only subsequently to amalgamate rather than to choose among them. Finally there was the practice of patterning: of arranging concepts in patterns of primarily aesthetic appeal, where the ordering principle might be classificatory, numerological, rhetorical, or built out of richly suggestive analogical correspondences.

Of these, the method of history was surest, since it dealt with a concrete subject matter which could be judged intuitively upon the basis of empirical observation. It was in the study of the historical record of "men and affairs" that traditional Chinese intellectuals were most solidly "rational." However, the second and third methods bypassed the whole question of any kind of verification and so opened the way for pure speculative thought — by which is meant thought relying neither upon empirical verification nor upon internal logical guidelines for its structure. Unrestricted by the claims of verification or any disciplines of logic, the speculations of Han syncretists, Ming metaphysicians, or students of the *I Ching* proceeded as a form of imaginative exercise; and a man's choice of one doctrine over another could ultimately be called a matter of personal preference, modified mostly, perhaps, by the weight given a doctrine by the *historical* tradition with which it came to be associated. Most Chinese theories about nature were arrived at by these speculative methods.

In late imperial times many Chinese sensed and were disturbed by the lack of methodological moorings in neo-Confucian metaphysical philosophy; empty words were the scholar's bane, and they were often blamed on the excesses of Buddhist theology, which indeed had suffused the Chinese metaphysical vocabulary after Sung times.

For those seeking the solid ground of verifiability, however, the most important Chinese philosophical school was that of the so-called "Ch'ing empiricists" of the early Manchu dynasty. Essentially a school of historical scholarship, it emphasized the philological analysis of texts in order to discover their true historical meaning, while committing its followers to a position of philosophical nominalism. Basically, then, the renunciation of speculation meant a return to history in the form of systematic critical study of the classical canon.[4]

In brief, the first Chinese exposed to Western science were inclined to some of the following intellectual predispositions: When thinking of nature, they imagined an organic, harmoniously functioning cosmos, where the social and moral order was integrally linked with natural processes. When thinking about thought itself, they deemed speculative methods to offer the only known way of talking about the universe at all, once men had departed from the authority of history or the certainties supplied by common sense empirical observation.

Technically, Western science first entered China with the Jesuits, who introduced some aspects of post-Galilean mathematics and astronomy during their long tenure as astronomers and calendrical experts at the Ming and Ch'ing courts. But the new subjects were little studied in a prosperous, self-contained society, where the comprehensive intellectual prestige of Confucianism was reinforced by its vast social power and where, on the observational level, the Copernican portrait of the solar system was only a novelty, and not in the least philosophically revolutionary.

Western science made its first permanent impact on China through the shock of technology — the ships and weapons which won the Opium wars, the machines and goods brought into treaty ports by victorious Europeans. The first real students of science in China were in fact students of technology — the pioneers in the fields of modern shipbuilding, metallurgy, and arms manufacture, who after 1865 were trained by foreign instructors at the Chinese government's experimental Kiangnan arsenal. Here the first important translations of scientific texts were made into literary Chinese.[5] By means of these and other, Western translations some scientific ideas began to trickle through to radical Chinese in the last decades of the nineteenth century. They were not accompanied by any real opportunities for men to study actual scientific disciplines. The Kiangnan arsenal and the similar Foochow arsenal always stressed applied rather than theoretical study, but even these institutions languished in the 1880s and 1890s for lack of official patronage. In those decades the classical system of education retained a tenacious, though dying, grip on people's minds: it dominated the schools and monopolized preferment, and its spokesmen successfully discouraged even pilot programs for study abroad. When curriculum oriented around Western subjects was at last authorized in officially sponsored schools for the general educated public, it was 1902. Even then, until a generation of trained Chinese professors had emerged equipped to teach scientific disciplines in their own language, advanced training in a scientific subject was possible only for the tiny handful who went abroad.

Soon, however, enthusiasm for science, which seemed to be one of the keys to Western civilization, far outpaced the slow development of scientific studies in the schools. Among early Chinese explorers of Western scientific learning, several characteristic patterns of discussion emerged. When presented with characterization of the structure of the universe associated with theories in physics or chemistry — say, contemporary molecular theory — they often assumed these were "facts" to be used as the building blocks for a revised speculative cosmology. For example, in 1898 the radical T'an Ssu-t'ung put his readings of available scientific translations to the service of a syncretic utopian metaphysic in which the physical concept of ether appeared as an all-encompassing cosmic force, expressible as "gravity" or "electricity," to be held responsible for the cohesiveness of the universe and its moral base. The cosmos that T'an envisaged was basically Buddhist and neo-Confucian in conception, and ideas drawn from European science were startling eclectic accretions onto a previously imagined structure.[6]

Further, because they assumed the fundamental unity of natural and moral truth, Chinese were quick to conclude that Western scientific theories were applicable to the world of society and had ethical relevance. The highly imaginative T'an claimed that goodness (*jen*) was a function of ether. But less avowedly visionary men, considering the popular subject of Darwinism, focused not on biology itself, but on the historical, social ideas it called to mind. The idea of evolution suggested a scientific base for a revision of Chinese attitudes toward time and change — the acceptance of the socially revolutionary idea of progress. The concept of natural selection gave rise to the new proposition that struggle and conflict are principles of social life, given free reign in the individualistic, competitive societies of the West, which Chinese civilization, to its cost, had tried to suppress.

Finally, Chinese soon learned that the basis for the enormous authority of science lay not merely in its technological possibilities, but in the presumed efficacy of the very special methods by which scientific research reached its conclusions. Given the decline of the Chinese historical tradition, along with the decay of Confucianism, and the long-suspected indiscipline of speculation, many radicals seized upon the idea of the scientific method with special zeal, as a new style of reasoning capable of providing a clear guide to decisions for an uncharted future. Their association of science with value made them receptive to the most sweeping claims of the then new social sciences and to identifications of the scientific method not only with certain procedures, but also with appropriate moral attitudes. In the words of a typical commentator, "The scientific spirit lies in rooting out dogmatic prejudice and submitting to objective truth . . . After men have undergone a scientific training, they can cultivate the virtues of carefulness, honesty and justice." [7] Beyond this, early descriptions of the scientific method were vague; many thought of it as a matter of simple, concrete observation followed by formal classification, echoing the empiricism of the Ch'ing philologists and the rhetorical formalism of literary tradition.

Such were the bases for the Chinese form of scientism — if scientism may be described as the application of scientific concepts to other, unrelated areas of inquiry outside their own sphere of relevance. Of course scientism was never peculiar to China; the intellectual attitudes sketched here all have had parallels in the West. Europe had its own prescientific traditions of speculative philosophy, its own theological cosmology, and a history of conflicts between these and the claims of science which resulted in the construction of scientistic metaphysical, sociological, and methodological systems on the largest possible scale. European positivism, beginning with Comte, was a movement dedicated to the systematic replacement of traditional Christianity and all forms of a priori philosophical and social thought with a broad body of doctrine derived solely through scientific verification. In Comte and other nineteenth-century thinkers like Spencer and Marx, "positive" philosophies took visionary forms, representing an extreme of scientistic system building, calling upon the propositions of science to justify the most comprehensive world-historical views. A later, contrasting group of positivist thinkers gave the movement a more critical turn. Men like John Stuart Mill, Ernst Mach, and John Dewey concentrated on investigating the epistemological and logical nature of the scientific method itself, and some of them attempted to construct models of this method broad enough to guarantee its applicability to a wide range of inquiry. Although a high level of scientific development was a precondition for profitable investigation of the issues surrounding science, the West's leadership in this field did not save Westerners from the dangers of scientism. On the contrary, it made scientistic speculation among intellectuals especially subtle and scientistic assumptions a more pervasive feature of popular culture. Scientism in China, then, was part of a wider phenomenon, and patterns in traditional Chinese thought explain something about the particular form of scientistic thought there, but do not account for its existence.

The Chinese seeking to understand science and bring it to the East had to maneuver his way as best he could among these complexities. On the one hand, a scientifically uncongenial native tradition supplied appealing preconceptions which made scientific theory opaque to him and scientific practice difficult. On the other hand, study in Europe exposed him not only to science itself, but to a whole range of shrill, conflicting interpretations of its meaning, offered with reference to the social and philosophical issues of an alien civilization, where, if clarity was possible, confusion was certainly commonplace. Meanwhile the ongoing technological explosion was carrying Europeans to ever expanding levels of material power, but in China society was sinking into a condition of revolutionary disorder. If the problems demanding understanding were formidable, the crisis demanding action was no less so.

Ting Wen-chiang's career was spent in the crosscurrents of these influences. Self-consciously he attempted to construct a comprehensive scientific outlook, capable of providing him with moorings in the controversial waters of European thought, of supplying the theoretical basis for a new, syncretic view of world civilization embracing both China and the West, and of serving as a practical guide to social reform at home. Ting has been called the Chinese Huxley. He offered the Darwinian world view — with its biological conception of man, its positivist view of knowledge, and its sober utilitarian ethic — as the answer to the twentieth-century problem of restructuring the Chinese intellectual heritage. Yet at the same time this scientific ethos, adapted by a man of Confucian gentry background to deal with the Republican crisis of social reorganization and state power, proved itself impotent to affect the course of political change.

Chapter II

The Returned Student

Ting Wen-chiang (V. K. Ting),[1] was the second of four sons born in a remote country town in Kiangsu in the last years of the nineteenth century. The village of Huang-chiao in Tahing hsien lies only twenty miles or so north of the Yangtze River and about a hundred and fifty miles northeast of Shanghai. In those years Huang-chiao was sleepy and conservative, little touched by the Western influences of the coast. As local gentry the Ting family enjoyed some connection with officialdom through relatives; Ting senior's grandfather had occupied a minor post in Chekiang and a close cousin was married to a well-known *chin-shih* of Soochow. But he himself was a man of purely local concerns, whom his sons remembered as occupied with village and clan duties. In keeping with family tradition, he was responsible for maintaining a free cemetery and other philanthropies for their kin and the local poor.[2]

The conventional biography of a scholarly Chinese usually skips quickly over childhood years, pausing only to make standard allusions to youthful precocity and perhaps to add an admiring aside concerning the mother, who taught the hero high ideals. In Ting's case such references, which are supplied by his brothers, may be taken as more or less accurate. What is certain is that young Ting, having been born in 1887, was a member of the last generation whose early education revolved around the requirements for the imperial examinations. Between the ages of five

17

and fifteen he spent long hours in a village school which groomed youths for the *sheng-yuan* examinations, and the available lists of his childhood reading indicate that he was steeped in literature designed to inculcate Confucian high-mindedness in aspiring literati. Predictably he read the *Three Kingdoms Romance*, Han Yü, Ssu-ma Kuang, Su Tung-p'o and the later Sung and Ming philosophers. Sometime before he left home at fifteen he also picked up an acquaintance with the writings of the early Ch'ing opponents of Manchu rule, Ku Yen-wu, Huang Tsung-hsi, and Wang Fu-chih. The fare was traditional, but it gave scope for personal taste. As a boy, Ting emphatically preferred the study of "men and affairs" to the mystical speculations of neo-Confucian philosophers or the classical philology of the Ch'ing scholars of "Han learning." Books on history and the lives of great statesmen were what attracted him, particularly those dealing with the controversies of the Northern Sung, when high scholar-officials debated fundamental issues of social and political reform.[3]

Such Confucian models allowed scope for resolute deeds, even heroism, and after the Hundred Days' Reform of 1898 first made him and his schoolfellows aware of China's dangerous internal weakness, Ting longed for the revival of such a militant spirit among educated youth. For a short time in 1898 the young emperor Kuang-hsu, surrounded by enthusiastic scholarly advisers, had tried to legislate a dramatic reform of an outworn bureaucratic structure. The schoolboys of Tahing, inspired by his example, briefly attempted their own revolt against antiquated routines. They swore that they would refuse to write any more classical compositions in the "eight-legged" essay style or waste hours practicing calligraphy. Instead they vowed to devote themselves to "solid studies" — the books of history and biography in which a man of traditional China expected to find the precedents and models for successful political action.

There is no way of knowing whether Ting's schoolyard rebellion was any more successful than that of the Kuang-hsu emperor against the conservatives at his imperial court. Ting, in any case, was graduated from his school in good standing two years later. Although he was eligible for the *sheng-yuan* examinations, he probably did not take them. His brother Wen-yüan recollected that the family was considering instead sending him to Shanghai, to the *Nan-yang chung-hsüeh,* one of the first Chinese secondary schools to offer courses in Western subjects. However, an event intervened which altered the course of Ting's life — he found a patron.[4]

The patron was Lung Yen-hsien, a student of the "new learning" (*hsin hsüeh*) from the West, who came from his native Changsha to be magistrate in Tahing in 1901. Having established a school in the district, he heard of Ting's academic promise. Summoning the boy, he examined him on the subject of the emperor Han Wu-ti's expedition against the southern and western barbarians. The results convinced Lung that Ting was destined for "high rank," and he knew that in the contemporary age the only adequate preparation for such a role lay in study abroad. Lung undertook the delicate task of persuading the family to send their fifteen-year-old son to school in Japan. The conservative gentry of Tahing were deeply suspicious of foreign travel and almost all counseled against it. There were personal obstacles as well. Ting's mother had recently died and his father felt the boy ought not to leave home during the family bereavement. Moreover, even a family of social position in China was rarely so prosperous that it could easily afford this kind of education for its children. For the Tings it meant that a loan would have to be con-

tracted. The selection of the most promising boy for educational preferment was a gamble on the entire family's
future and a necessary curtailment on the opportunities
of others. The oldest son, Ting Wen-t'ao, wanted to study
abroad too. However, the more ambitious and brilliant
younger brother was determined that the opportunity
should be his. Wen-t'ao wrote of it later, "My brother said
to me, 'If no one stays at home, who will take care of
the family? If no one goes abroad, who will plan for the
nation? Family and nation, you and I will divide the
responsibility.' " And Wen-t'ao added rather forlornly,
"Thereafter I was willing to retire to accomplish my brother's goals." [5]

Ting never forgot this struggle for emancipation. "If I
had not happened to meet Lung Yen-hsien," he told a
friend long afterward, "the entire course of my life might
have been different." [6] A few years later he removed his
favorite younger brother, Wen-yüan, from Tahing hsien,
saying it was not a suitable environment for the upbringing
of an educated man. Shortly before his death, when he had
occasion to think back over his early schooling, Ting remembered most of all its inadequacies. He remarked upon
the generally low quality of the old scholarship even among
those of literary reputation. From the perspective of the
1930s it seemed to him that his former teacher, who held
the prestigious rank of Imperial Licentiate (kung-sheng),
wrote letters with less facility than did the average Republican upper-middle-school graduate. Above all he criticized
the old scholarship's total disregard for natural observation and the text-centered curriculum which failed to prepare students for any kind of direct empirical reasoning
about the physical world. Children of the gentry were not
even encouraged to participate in physical activity which
might bring them into intimate contact with nature, so

that "literary feebleness" (wen jo) was a common mark of an educated man. Before he left China at fifteen, Ting had never walked farther than a mile at one time. When he arrived in Japan, even a class in geometry, whose subject matter is points and lines, seemed absurd to him.[7]

However completely Ting may have abandoned the intellectual habits of his childhood education, its moral imperatives were not so easily repudiated. To the end of his life he bore the complex of obligations with which he had left home at fifteen. He was the custodian of the family honor and perhaps of its fortune, first among a swarm of younger brothers (soon augmented by his father's remarriage), who, in a clan with a strong tradition of mutual aid, would eventually look to him for the furtherance of their welfare. He was a member of a privileged class, for whom the advantage of birth was being enhanced by a final refinement of training, bringing with it the expectation of government service and perhaps the duty of leadership in the revival of his country. The firm sense of hierarchy imparted by his family and village environment remained with him as a lifelong sense that his custodial role was natural. Neither did he ever forget the moral lessons of his Confucian home, where leadership and scholarship were inextricably linked and the authority of a man of affairs was presumed ultimately to be justified by his personal virtues.

Ting stayed in Japan from the fall of 1902 until the late spring of 1904. Once he was plunged into the hectic world of the Chinese students' ghetto in Tokyo his formal studies soon took second place to politics. This was before Liang Ch'i-ch'ao and Sun Yat-sen had entered into a serious competition among overseas students for their allegiance to the alternative goals of constitutional monarchy or a republic. Ting easily combined enthusiasm for Liang's writings with vaguely antidynastic and Republican sentiments which he articulated in essays composed for the Kiangsu students' paper. At practically no expense the students were able to publish voluminously. Ting later recalled that at one time no more than a thousand Chinese students were turning out six monthly reviews of two hundred pages each. These publications were transmitting back to the mainland the radical ideas of what was already beginning to be named the "Chinese Renaissance." In addition to demanding political reform, the student innovators preached against opium and footbinding, ridiculed classical Confucian education, and recommended the emancipation of women. Inspired by Japanese linguistic example, they tried also to encourage the use of a more modern literary style dedicated to greater intelligibility, ease, and freedom of self-expression.[8]

Equally important, student life in Japan encouraged the development of a more self-conscious Chinese patriotism among young men who were cut loose from their provincial home districts and able for the first time to savor the sense of common national identity fostered by living in a somewhat hostile and indifferent foreign country. Many of Ting's friends in Tokyo were military students. Because the Taiping rebellion had put military leadership in the hands of Chinese rather than their nomad rulers and a series of disastrous wars had made "self-strengthening" a matter of urgency, prejudice against the military profession had begun to decline among educated Chinese. Encouraged by the government, military studies had priority for students in Japan, and Ting's developing sense of patriotism, like that of the cadets, had a decidedly martial ring to it. For a time he even toyed with the idea of a military career. His decision to leave Japan was hastened by the outbreak in February 1904 of the Russo-Japanese

War, which alternately filled the Chinese student community with shame and pride: pride that an Asian people were capable of trouncing a Great Power and shame at Chinese impotence compared with the glory being won by the once despised islanders. Ting and his friends smarted under the taunts flung at them by Japanese they met in the streets, and Ting's original ambition when he left for England was to prepare himself for a naval career.[9]

The idea of going to England was first suggested by a letter from Wu Chih-hui which circulated among Ting's friends. A revolutionary anarchist and later a follower of Sun Yat-sen, Wu in 1904 was already notorious, having recently been deported from both China and Japan for anti-Manchu agitation. His letter came from Scotland, where he was passing his time in exile studying evolution and paleontology. "Chinese students in Japan go to meetings in the evening, eat Chinese food, talk politics and don't study,"[10] Wu wrote scornfully; he estimated that a Chinese could study authentic Western learning in Edinburgh on an income of six hundred yuan a year. Gradually three friends, Ting, Li Tsu-hung, and Chuang Wen-ya, determined to try to reach Europe. Even though he was the youngest of the three and the only one who knew no English, Ting took the lead in their plans. He persuaded Li's family and his own to consent to their scheme, and he bought the steamer tickets. In order to calm the anxieties of his companions, he argued that they could stretch their funds by placing them in a common pool: by the time it was exhausted they would be able to write home for more. In the late spring of 1904 they departed, aboard a German steamer because the cheaper Japanese ships had been canceled as a result of the war. Their financial provisions were worse than inadequate. After paying their ocean fares and other bills incidental to departure, they embarked for Europe with less than fifteen pounds gold among them. They were so ignorant of Great Britain that they did not even know that Edinburgh was a long train journey from London.

Happily they were saved from the possible consequences of their recklessness. In spite of their antidynastic views, on a stopover in Penang they paid a courtesy visit to K'ang Yu-wei in exile. The great man honored them with a "discourse of encouragement and warning to youth" [11] and also perspicaciously suggested that they accept a loan of ten pounds and an introduction to his son-in-law in London. These aids helped them reach Scotland, where they faced a more serious crisis. Wu Chih-hui met them in Edinburgh, but they quickly found that his financial calculations had been overoptimistic and that Wu himself was penniless. In the end Chuang Wen-ya, who was the poorest of the three young men, accompanied Wu to Glasgow, where they planned to work as dock laborers; Ting and Li were saved by the kindly intervention of another stranger. They were befriended by a certain Dr. Smith (*Ssu-mi-shih*), a former medical missionary in China. Realizing that they were both poor and entirely unprepared for the academic tests of British higher education, Dr. Smith suggested that they attend secondary school in his own home town of Spaulding in Lincolnshire.

For most of 1905 and 1906 the two Chinese lived inexpensively in this modest provincial town, where their sponsorship by a respected citizen made them generally welcomed. Here Ting gained his real knowledge of English life. He later called Spaulding his "second home," and on his last trip to Europe in 1933 he went back there to visit his friends of thirty years earlier. [12] In the local school he received his first serious introduction to Western learning: mathematics, history, geography, physics, chemistry, Latin, and French. In two years he progressed through four forms and achieved the English schoolboy's highest prize, a place at Cambridge.

But Edwardian Cambridge was not the nursemaid of "lesser breeds without the law." Ting, although he was by now a Chinese government scholarship student, found the style of life there quite beyond his reach. At the end of a single term he withdrew, and after a few months on the Continent, he transferred in the spring of 1907 to the more modest atmosphere of a Glasgow technical college. After a brief flirtation with the idea of becoming a doctor (ended by a refusal of admission to the University of London), he settled down to the serious study of zoology and geology at Glasgow University. These experiences could not help but bring to his notice the classical bias of upper class education in a nation supposedly at the forefront of industrial progress, and they contributed to his later understanding — far more sophisticated than that of most Chinese — of the differences between technology, which men easily learned to manipulate, and scientific principles, which often remained remote from their understanding.[13]

Ting received his B.S. in 1911 with a double major in zoology and geology. During vacations he traveled extensively on the Continent, particularly in the Alps, and improved his French and German. He left for home in April of 1911 after an absence of seven years.

Ting reached England as a young adolescent, and he left as a grown man whose character had been deeply influenced during the intervening years. From a boy full of incoherent revolutionary sentiment and a commitment to "save China with study," he had developed into a professionally minded adult with an ambition focused purposefully upon the scientific modernization of his country — both technical and intellectual. "A most deeply Europeanized Chinese," Hu Shih said of him.[14] Westerners often reacted to his personality with a sense of relief at meeting something familiar. "The ablest man I met in China," said Bertrand Russell;[15] and Teilhard de Chardin responded to him with the reflection, "Perhaps there really is a 'new China.' "[16] On the other side, Ting's intimacy with foreign scientists, his versatility in languages, his forthright manner of speech, even his mustache and his liking for cigars, were occasionally disconcerting to the more old-fashioned Chinese. His friends spoke reassuringly of his "essential" Chineseness, and those who were not his friends deplored his deracination. Actually the true basis of all these reactions was that in Britain, Ting had mastered the international vocation and intellectual style of the scientist.

In science Ting found the intellectual mistress of his life. On the practical level it gave him important work to do. Like most Chinese, he originally had been drawn toward science by its utility. Geology was directly applicable to the development of China's mining industries, which in the early twentieth century remained either neglected or under foreign control, and Ting was destined to be a rare Chinese technical expert in this field. But more important, in Edwardian England the biological view of man still retained some of the potency of a recently discovered and controversial truth of such broad explanatory force that much in man's idea of the world was inevitably affected by it. The most eminent proponent of Darwinian science, Thomas Huxley, was also the chief intellectual influence upon Ting during his stay in England. Stimulated by the ideas of the great Victorian, Ting found in science a world view, an intellectual methodology, and a touchstone for value.

The fundamental axiom of this world view was methodological: scientific reasoning provides the sole guide to truth in all matters about which human beings may reliably know anything. By this basic positivistic tenet, nineteenth-century scientific rationalists hoped to repudiate the claims to authority of all forms of speculative and metaphysical thought as well as the claims of political and social beliefs which enjoyed acceptance solely because of traditional sanctions. It was obvious that scientific reasoning can only be applied to the natural world — that is, the world of perceptible, empirically verifiable entities. Accordingly scientists reduced the scope of what could be considered capable of yielding genuine knowledge to the world of nature, but at the same time they expanded their idea of that world to embrace man himself, the biological animal and his social institutions. This comprehensive naturalism received from its opponents the opprobrious epithet of "materialism" since it treated human beings as if they were governed by the same scientific rules that govern the behavior of all living matter. In Europe the new materialistic psychology seemed most subversive to defenders of the Christian religion. When scientists declared war on all "spontaneity," by which they meant the idea that a phenomenon could possibly lack a cause explainable in terms of empirically observable events, they suggested to beleaguered theologians that psychic phenomena too must be explained by physical causes, so that prevailing theological beliefs in spiritual existence and in free will were made to seem untenable. Although Huxley enjoyed teasing his adversaries with disquisitions on the implications of physiological psychology, he was not at heart a philosophical materialist. Rather, because of a deep-seated skepticism, he consciously limited the range of his intellectual inquiry to the perceptible "real" world. He himself coined the word for his position, "agnosticism," and it placed him in the camp of the critical, as opposed to the systemic, positivists. According to Huxley, since scientific methodology provides the only reliable guide to human understanding, human beings are limited in what they may know by what can be tested and verified by its procedures. Whatever psychic reality may be, men can only talk about it in "material" language.

Huxley's religious skepticism was easily assumed by Ting, who had submitted to the clerical indoctrination of his Anglican English schools with the polite indifference of a Chinese nurtured in a primarily ethical Confucianism. The antisuperstitious, empiricist temper of certain strands in Confucian writings also helped him to relate this skepticism to his native intellectual tradition and to interpret European religious ideas as only one form of a range of metaphysical and speculative beliefs, many of which in Buddhist, Taoist, or neo-Confucian guises were obscurantist influences in the China of his time. The basic axiom of his thought came to be the demand that all kinds of traditional ideas, Chinese and Western, be scrutinized solely with an eye to their scientific verifiability.

To nineteenth-century British scientists like Huxley, and to Ting after them, scientific verification was fundamentally inductive in method and broadly sociological in scope. They thought of facts in terms of incontrovertible concrete observations and they thought of method in the terms outlined by J. S. Mill's *Logic*. Scientific reasoning seemed to them a basically empirical, inductive enterprise that moved from fact to generalizations based upon classifications of facts, from what was clearly perceptible described in the language of ordinary experience, to what had the status of scientific law. Such a notion of scientific procedures made them optimistic about the claims of scientific sociology. Once it had been shown that man the animal lives under the same laws that operate for all organic nature, there was every reason to extend the investigation to seek general rules covering the behavior of man the social being and thinker. Inductive classification of discrete facts and generalizations concerning them seemed a method for the analysis of psychological behavior (psychology) or social customs (anthropology and sociology) just as practicable as it had proved to be for the analysis of the botanical or paleontological record. The scientific positivist was convinced that the solution to social problems lay in the scientific analysis of institutions just as the solution to the riddle of the earth's history lay in the analysis of the geological record.

Moreover, conclusions in sociology were imagined as enjoying the same autonomy as did conclusions in the physical sciences, arising out of experimental data without any interference on the part of human presuppositions, and producing results which men had no rational choice but to accept. It was a notion of scientific method which did not take into account either the theoretical component of scientific reasoning or the interpretive latitude with which almost any experiment can be constructed and analyzed, and it produced a quite simple view of how scientific solutions could be provided for social problems. In considering a social problem, Ting, like his mentors, assumed that when the facts were in and the scientific analysis had been made, problems of interpretation had been solved. Scientific conclusions about the state of society were taken as automatically providing answers to questions about social choices. In practice this made Ting the scientist hostile to political or social ideology — a pragmatist in political action.

The commitment to accept as the highest truth that men may know the blunt intractable conclusions of scientific reasoning forced Victorian scientists into a somewhat bleak ethical stance. Since they believed scientific fact to be a kind of ultimate reality which men must accept as true and to which they must conform, in theory facts replaced values as guides to human choice and conduct. Man's biological animality, his temporal mortality, and the finitude of his planet were physical facts which he must not merely accept but honor as among the great truths of human life. Out of the necessity to identify fact and value arose the contradictory teachings of social Darwinism, which left its mark on both Huxley and Ting. Behind all versions of social Darwinism lies the impulse to describe admired types of human conduct, whether the heroism of acts of competitive struggle or the altruism of those of social cooperation and self-sacrifice, as facts of nature. In both Ting and Huxley there existed a life-long tension between their experimental standards, which taught them to deny any imputations of teleology to evolution, and the temptation, when pressed, to justify certain social and ethical ideals which they valued by describing them as the biologically conditioned products of natural selection.

By the time he was ready to return to China, Ting had developed a set of views on science which claimed the broadest relevance to China's problems. The belief that rational reform of social institutions requires only the application of scientific reasoning to the problems of human organization and welfare had been proffered in a rather tentative, visionary manner by individuals sheltered by the stable verities of life in Victorian England. But it assumed a more dogmatic, programmatic form in a man like Ting, for whom the spectacle of China's intolerable backwardness and disorganization was agonizingly immediate. In the application of scientific methodology he thought he saw a standard of judgment for the elimination of antiquated customs, superstitions, and folkways which stood as obstacles to national modernization, and he hoped to find guidelines to the programs which would shape Chinese society in the future.

At the same time the Darwinian world view eased the sting of China's backwardness by making it possible to understand her society in a global historical framework. It made a truly international perspective tenable for Ting and helped him to reconcile Chinese tradition with Western power. On a planet where land masses and oceans had taken eons to assume their present form, where the earth's bedrock and the fossil record revealed a common global evolution, human beings themselves had achieved their recognized shape only recently and had spread out over the earth in slow, pulsing migrations, leaving signs of a common paleolithic and neolithic past. In such a world the cultural gap between China and the West was trivial: the autocracies which collapsed in Europe a hundred years before would fall tomorrow in Asia; the technology which Englishmen gloried in today would soon become the common property of mankind. England, like China, had her classical past: Christian mythology and Greek and Latin education, which the scientists of the nineteenth century fought to eliminate from British education and society just as the Chinese of the twentieth century were fighting Confucian reaction. In scientific thought lay the sole criterion of truth, universally applicable to mankind just as physical laws rest upon the uniformity of nature. Out of the seemingly irreconcilable clash of cultures and creeds, free scientific thought would eventually emerge triumphant and organize itself into a coherent system embracing the whole world.

In this vast historical framework the vocation of the scientist was a high, even a sacred, calling. He was the custodian of the knowledge upon which all progress in human societies depended. The pursuit of that knowledge required of him an unflinching integrity in the fact-finding and truth-telling operations which were the essence of his experimental labors. Huxley had viewed his scientific work with an idealism which aroused Ting's eager admiration; no mere technical competence could have satisfied the lofty sense of vocation instilled in Ting by his Confucian conviction that social and moral purposes should be firmly linked to any worthwhile intellectual activity. Through science Ting hoped to satisfy his need for a career which would revolve around an intellectual enterprise of unique moral seriousness and comprehensive relevance to human affairs.

When Ting left for home in May 1911, his training had prepared him to undertake a wide range of stimulating work. Inside China geology, paleontology, archaeology, and anthropology were literally unstudied; the potential contribution of Chinese data to global issues in these earth sciences remained to be guessed at. Even the country's climate, topography, and population had never been precisely mapped. In early twentieth-century Europe geology was a mature specialty whose researchers worked within strict boundaries, but the position of China as terra incognita gave Ting some of the catholicity of classical nineteenth-century naturalists. Huxley had dabbled in observations of the aborigines of New Guinea, Darwin had studied barnacles in Sussex and fossil man in Patagonia, and Hooker had explored Antarctica. Similarly Ting spread his net wide, traveling over all China, doing cartography in one place, physical anthropology in another, and in a third combining exploration and mineral prospecting with studies of stratigraphy and the paleontological record.

During the first years after his return, Ting devoted himself entirely to teaching, exploration, and research. In spite of the comprehensiveness of his scientific ideal, in the beginning he remained a specialist, bounded by laboratory and classroom. After 1919, when he did concern himself actively with a scientistic program for political and social reform, he did so much in the spirit of Western positivism: rather than using the advocacy of science only as a propaganda weapon against reaction, he also sought ways by which scientific procedures might be made to yield in history, politics, and economics results as stable as they had provided for physics and biology. In this respect, Ting's European education had made him Westernized, not just a Westernizer.

Chapter III

Odyssey of a Chinese Scientist

The journey back to China in the summer of 1911 offered Ting his first opportunity for scientific investigation. He arranged to leave his ship in Indochina and to take the Yunnan railway, then recently completed, up to Kunming. From there he followed the main imperial courier route through Yunnan, Kweichow, and Hunan, traveling by foot and river junk through a remote mountainous terrain of great geological and mineral interest, making cartographical and geological observations as he went. He arrived home in Kiangsu just two months before the outbreak of the insurrection against the Manchu garrison at Wuchang in October 1911.[1]

Almost without visible leadership, rebellion spread in every important province, and within a few weeks it became clear that the Manchu empire was irremediably damaged. Yet the fall of the dynasty, so thrilling a possibility to Ting as a student in Japan seven years earlier, at first made little immediate difference to his private or professional routines. He spent the revolutionary year of 1911–1912 as a schoolmaster in Shanghai, teaching science to the secondary students of the *Nan-yang chung hsüeh*. He also acquired a wife. His marriage, to Shih Chiu-yüan of Soochow, was clearly a suitable family alliance with a lady who was described by his friends as charming and even educated, but whose ideas were not always of his world. In his married life, which people said was entirely

34

happy, Ting accepted traditional practice with one excep-
tion — the union remained childless.[2]

Concerning the revolution itself, Ting assumed a
simple traditional gentry role of local leadership in a time
of disorder: he returned briefly to Tahing hsien to organize a village defense brigade against possible incursions
by bandits or leaderless soldiers of either army. When two
friends who were in Nanking on the staff of the Republican army commandant, Hsü Ku-ch'ing, urged him to become one of Hsü's secretaries (*mi shu*), he refused, declaring that a man of his training could do most for the revolution by fostering science and modern industry.[3]

However, the establishment of the republic led to a
decisive professional opportunity for Ting. In the winter
of 1911–1912 it was decided in Nanking to create a geology department under the Bureau of Mines (*K'uang cheng
ssu*) of the new Ministry of Commerce and Industry
(*Shang kung pu*). The following year, after the ministry
had moved to Peking under the united Republican administration, Ting was invited to take charge.[4] In this way the
revolution opened the path for him to undertake the
enormous labor of creating an environment where the
modern study of the geological sciences would be possible
in China by Chinese. Under Ting's leadership the geology
department grew into the Geological Survey of China
(*Chung-kuo ti-chih t'iao-ch'a so*). He was its director until
1921 and remained deeply involved in all of its enterprises until his death.

In 1913 the problems of undertaking geological research in China were formidable. From his first weeks
back home Ting had begun to gain practical experience
of the difficulties before him. For example, even as he
traveled through the southwest in July 1911, he learned
of the inadequacy of existing maps of China. At that time

atlases copied universally by Chinese and Westerners alike were still based on the charts prepared for the K'ang-hsi Emperor by the Jesuits in the seventeenth century. Ting had in his possession a widely used atlas published by the Wuchang Geographical Society (*Wu-chang yü-ti hsüeh-hui*) under the Ch'ien-lung Emperor and a recent map from the Commercial Press of Shanghai; both relied heavily upon the K'ang-hsi model.[5] Not only did these maps lack topographical data, leaving basic features of the physical geography of inaccessible regions obscure, but they did not even accurately reproduce the path of the imperial courier route Ting was traveling. A change in the route subsequent to the K'ang-hsi period had remained unnoticed down to recent reprintings.[6] Twenty years afterward Ting coauthored China's first complete modern atlas, based in part upon the cartographical measurements he had accumulated during that first trip.

As a young schoolmaster in Shanghai, Ting had seen personally the inadequacies of scientific training in the classroom. Although the *Nan-yang chung hsüeh* was considered a very up-to-date institution, he could give his students only the most elementary and superficial introduction to general science. His classes were expected to cover the subjects of biology, zoology, geology, chemistry, and Western history, yet lectures were taken up explaining the simplest concepts, like that of "erosion." Good schoolbooks in Chinese were hard to find; his own zoology lecture notes were later expanded into a middle-school text.[7] Ting soon learned that conditions in secondary school were perpetuated in college. Peking University, which was heir to the old Manchu *T'ung wen kuan*, the special imperial college for training diplomatic personnel, and could be considered pre-eminent in Western studies, was not offering courses in geology in 1913. A German

professor, Herr Solger, had been imported for the purpose
in 1910, but his classes had been discontinued for lack of
students.[8]

Ting noticed these visible practical deficiencies in the
scientific environment in China; behind them lay a complex of intellectual misconceptions, from the Confucian
prejudice of traditionalists to the scientistic fantasies of
radical youth and the narrow utilitarianism of "practical"
men.

Under the early republic the most serious intellectual
obstacle to science remained the legacy of Confucianism,
which could not be erased in a decade or two of interest
in Western studies. The perceptions even of an early geology enthusiast had been deeply colored by this humanistic
Confucian background. Hua Heng-fang of the Kiangnan
arsenal had translated Lyell's *Principles of Geology* into
Chinese as early as 1872. But he confessed at the time he
knew nothing of the subject: the concepts it treated were
so difficult to grasp, he said, that for months during the
work his mind was haunted in dreams by images of the
fantastic prehistoric animals Lyell described.[9] Lyell's
work in natural history was most easily interpreted even
by its translator in terms of fantasy and myth.[10] Predictably it found few readers among conventional Chinese of
the late nineteenth century, to whom the occasional Western naturalists who undertook field trips in China appeared engaged in an eccentric pursuit unworthy of a
scholar's dignity. Students of the early twentieth century
had not yet broken away from these attitudes; if they approached the study of geology at all, they tended at first to
be indifferent to natural observation and to dislike the
physical labor involved.

On the other hand, the vogue of Darwinism which had
made terms like "struggle for existence" and "survival of

the fittest" popular among radical Chinese youth had, in fact, little to do with the intellectual discipline of science. Rather, science enjoyed vague prestige as the source of Western power, and this prestige led young Chinese to suppose for a while that in Darwinism they had found intimations of the moral order upon which this new world of power was based. They preferred the clichés of strife and contest, which seemed to sum up the spirit of Western imperialism, or else they turned to the sentimental humanitarianism of Kropotkin, which made evolution into a saga of developing social cooperation and mutual aid; or to the Huxley of *Evolution and Ethics,* in which Confucians, like Christians, could find comfort in the proposition that the process of natural selection has led directly to the creation of ethical customs, and that these in turn are in the process of modifying the future course of human evolution in a manner which will ensure their own preservation and refinement. Morality — either the selfish morality of power or the altruistic one of cooperation — is justified by natural history as by human tradition. This was the lesson Darwinism taught the reform-minded Chinese youth. But nowhere had they learned from it to consider science a rational method for ordering and interpreting observations about the world.

If traditionalists and reformers alike responded to the idea of science with philosophical or literary reflections, practical men often continued to see it according to the view embodied in the *ti-yung* slogan — as a purely technological matter, a mechanical trick for self-strengthening. Such attitudes worked to the disadvantage of scientific study when they encouraged a narrow academic emphasis on application at the expense of principle (a weakness of the nineteenth-century arsenal schools), or when they reinforced the lingering social prejudice of educated

Chinese against "mechanical" occupations. However, interest in industrialization and the need to develop native technical experts was also the opening wedge which stimulated official support of science. In ministerial eyes the duty of Ting and his new geology department was to create specialists who could challenge the foreign monopoly of modern mining operations.

In February 1913 Ting arrived in Peking at the geology department's offices, which had been opened shortly before his appointment. He found that they consisted of one large room, devoid of books, maps, or specimens, and of three assistants so unlettered in the field that they did not even know of the existence of famous coal fields in the Western Hills near Peking. In practice the section was a bureau for processing official documents. Appropriations were scanty, and the only way to finance field work was through merchants who asked the department for aid in prospecting. Only one other Chinese with advanced geological training could be found to join the section — Chang Hung-chao, a graduate of the University of Tokyo.[11]

For a practical guide in the development of geological research Ting and Chang were forced to turn to the writings of a Westerner. The first real precursor of modern geology in China had been a German, Ferdinand von Richthofen, who, in a series of seven journeys beginning in 1868, traveled through large sections of the empire and constructed a rough, speculative outline of its geological structure. His findings appeared in five volumes, supplemented by an atlas, and although the work was less complete than the author wished, it remained the standard, indeed the only, reference for thirty years. When Ting and his associates began to pick up the trail of geological research after 1913, they found that Richthofen had saved them years of preliminary labor. However, Richthofen had made one comment about geology and the Chinese which served more as a challenge than as an inspiration: ". . . the Chinese man of letters is sluggish and chronically loath to move rapidly; in most cases he simultaneously vexes one with his avarice and cannot free himself from native prejudices concerning decorum. In his view to go on foot is demeaning, and the occupation of geologist a direct surrender of all dignity in the eyes of the world." [12] Ting felt this aspersion so keenly that seven years later he placed Richthofen's quotation at the head of his introduction to its visible refutation, the first issue of the first all-Chinese technical journal in geology.[13]

It was obvious to Ting and Chang that their immediate task was to train geology students. Fortunately the moribund Peking University geology department could still be of some use. Ting persuaded the academic authorities to lend him their small stock of equipment, to supply dormitory space for students, and to permit the use of Herr Solger as a lecturer in the new school. Chang I-ou, head of the Bureau of Mines, promised a subvention.

Winning over Solger was a more delicate matter, but Ting set about it with characteristic determination not to allow national differences to stand as obstructions to scientific cooperation. The German professor, who had passed a misanthropic three years on extremely bad terms with his colleagues at Peking, found himself invited on a field trip by the younger Chinese geologist and coaxed into friendship. Respect for his scientific work and a willingness to learn were the ingredients which thawed Solger's reserve, and Ting supplied the moral for his compatriots: "It can be seen that when foreign specialists cannot cooperate with Chinese, it is not always the foreigner's fault." [14] In 1914 another idle foreign expert was drawn into the department's work. He was J. G. Andersson, former director of the Swedish Geological Survey, who had come to China with two assistants to act as mining adviser to a ministry which had little notion of how to employ his time. Although Solger left to join the German army at the outbreak of World War I and Andersson was an official staff member for only one year, the pattern of Sino-foreign cooperation had been set.

In this way Ting and his organization began to overcome one of the standing early obstacles to the effective pursuit of scientific research — the psychological handicap imposed by the monopoly of the foreign expert. In the classroom even the best Western instructors were limited in their ability to communicate a subject. They were also an ambiguous, indeed distasteful, scholarly model for young Chinese considering whether to take up an unconventional course of study and brought up to want in a professor a personal patron and, if possible, a moral inspiration. When geology was at last taught in Peking by competent Chinese, then and only then did student enrollments begin to rise. At the same time it was important that feelings of national pride not prevent the Chinese from making the fullest use of foreign research talent. In geological and related circles there developed an unusually free atmosphere of cooperation between Chinese and Western specialists. To a considerable measure this was attributable directly to Ting's example of personal openness.

The geology school of the Bureau of Mines, known officially as the Geological Institute (*Ti-chih yen-chiu-so*), opened in 1913, and in the summer of 1916 it graduated its first class of thirty students. Ting, Chang, and Solger did most of the teaching. In 1914 Solger's place was taken by Wong Wen-hao (W. H. Wong), a recently returned graduate in geology from Louvain in Belgium, who soon became one of Ting's closest lifelong colleagues and friends. The tiny staff was forced to spread itself over a wide range of specialties and occasionally to double up on teaching loads. But of the class that was graduated in 1916, eighteen were immediately hired as junior researchers, while several of the ablest were sent abroad for graduate work.[15]

Duties at the Geological Institute were frequently interrupted by field trips, especially by Ting. In December 1913 and January 1914 he spent about six weeks in Shansi with Solger, investigating the Taihang Mountains and the iron and coal deposits of the Ching-hsing and Yangchuan districts. He had been back in Peking no more than a few days when he received a second and even more important assignment, to go to Yunnan; the subsequent trip stretched itself out to almost exactly a year of wide-ranging exploration. The official goal of each trip was mineral prospecting — in Shansi along the Cheng-tai railroad, and in Yunnan in the eastern section of the province, where the Ministry of Communications and the Sino-French Industrial Bank had just made an agreement concerning a proposed railroad linking Kunming and Chungking. But in each case Ting used his opportunity to range over a variety of studies. The results were typical of his methods of scholarship and of their particular value: pioneering maps and surveys were made in a number of fields; specimens were collected which formed the basis of monographs later written by others and also served as exhibits in the department's museum; sites were located which were likely to reward closer study later on. In addition Ting traveled as a Chinese with a lively interest in unexplored facets of his country's history. He visited some of the most famous mining centers of the old empire — the

Ching-hsing coal fields and Yangchuan iron workings of Shansi, and the Kokiu tin mines and the Tungchuan copper mines of Yunnan, and he collected materials for a history of the Chinese mining industry. In Yunnan he began to study the physical anthropology of the aboriginal peoples of the southwest, to map aboriginal dialects, and to collect Lolo religious inscriptions. On both trips he kept a spirited travel diary, full of observations on local people and customs, descriptions of the physical geography of the landscape, and anecdotes, often wryly humorous, of the ever present hardships of the road.[16]

The trip to Yunnan was the more memorable, both scientifically and personally. Ting loved exploring, and from his first trip there in 1911, the southwest of China held special attractions for him. From a geological point of view Yunnan in 1914 was almost completely unexplored. Richthofen had not penetrated deep into the region, and the only worthwhile modern observations came from a short visit by the French geologist Deprat in 1909. Even the basic outlines of the physical geography of the complex Tibetan foothills bordering eastern Yunnan were known only because of the travels in the seventeenth century of a Chinese explorer, Hsü Hsia-k'o. Hsü had discovered that the Chinsha River, which forms the border between Szechuan and Yunnan for several hundred miles, actually is the major source of the Yangtze. Ting's observations in Yunnan laid the foundation of modern knowledge of the region's structural geology. In geography he filled in the outlines supplied by his predecessor Hsü, and he faithfully retraced the path of Hsü's long-ago journey, partly as a romantic gesture of homage to the man he admired as a "patron saint" and partly so that his corroboration of Hsü's observations might bring to the Chinese geographer a long-deserved posthumous fame.[17] Finally, in paleontology Ting collected specimens numbering in the thousands which formed the raw material for a number of monographs later written by his colleagues in Peking for the journal *Paleontologica Sinica*.[18]

Moreover, this trip was one of the happiest adventures of his life. Compared with the wild grandeur of the Yunnan mountains, a traditional beauty spot like Hangchow seemed insipid to Ting. He had a true naturalist's taste for vigorous outdoor pursuits, spiced by special awareness that feats of physical hardihood flouted the traditions of his class and exemplified the spirit which was needed if a new generation of Chinese was to grasp the possibilities of modernization. He traveled as a Western explorer would, with tent and camping equipment to supplement rural lodging houses and with scrupulous attention to hygienic safeguards against vermin, contaminated water, and other health hazards. He took special pleasure in the amazement of porters and local villagers at this visiting "official" who neglected his sedan chair to scramble up and down mountains on foot and who, in the evening, was likely to ask for a bath or even to go swimming.

Ting's first stop in Yunnan was Kokiu, site of one of the world's biggest deposits of tin and of the largest export mining operation in China at that time. When Ting visited Kokiu in 1914 the district was exporting up to ten thousand tons a year at a value of sixteen to twenty million yuan. However, he found that 95 percent of the work was still being done by hand extraction methods, employing over twenty thousand men, many of whom were peasants who traveled to Kokiu for work during the slack season. The deposits were located high up in the range of limestone mountains which form the watershed between the Red River and the West River basins, under conditions which made traditional mining techniques more than usually effective. The ore was found in the form of pebbles embedded in veins of red clay and was easily extracted. The porous limestone of the mountains provided natural ventilation in the narrow earth tunnels, and the altitude reduced the danger of water damage. Laborers separated the metal from the earth entirely by the action of water, which was ingeniously channeled and stored during the natural rainy season; and the plentiful supply of men placed no obstacles in the way of sifting the sludge up to ten times. To smelt the ore, artisans used primitive funnel-shaped furnaces, stoked with charcoal and intricately regulated by hand-operated bellows.

Ting was compelled to admire the ingenuity involved but he deplored the way traditional techniques wasted usable ore and the cost in human toil. Although it was obvious that future development of the mines would be impossible without modernization, yet in 1914 the few pieces of European equipment which had been imported to the site — chiefly a mechanical sifter and a small cable railway — lay unused for lack of skilled management.

Ting envisioned the introduction of modern machinery as a means to liberate workers from wracking labor rather than as an innovation which would deprive peasants of badly needed marginal income. A day spent in the tunnels and a night passed in the men's barracks filled him with indignation over the miner's life. In the tunnels, which were often no more than two or three feet high, the work was hard, hot, and dangerous. At night the men were housed in filthy, overcrowded quarters, fed on rice and bean soup, and supplied with a water ration so meager that their choice lay between washing and having something to drink. They dressed in rags and were chronically ill; from continued contact with the ore their skins gradually took on a permanent reddish hue. Yet to his surprise, jobs at the mine were much in demand and sought by men who came from several days' journey away. On the trail a few months later, about a hundred miles to the northeast, Ting came upon a band of ragged peasant travelers whose stained bodies bore the marks of employment at Kokiu. They told him that because "the foreigners have started to attack one another," a production shutdown had occurred. Dismissed without wages, they were making their way home, kept from starvation only by the charity of the countryside and what they could steal. In this way Ting first learned that there was a great war in Europe.[19]

Beyond Kokiu, Ting pushed ahead into the most rugged and adventurous portion of his journey — through the little-traveled mountains near the Tibetan border, a region which contains some of the highest peaks in China proper and through which runs the remote and breathtaking Chinsha River. The complexities of the watershed and the romance of the landscape are both derived from the fact that the Chinsha and its tributaries cut deep canyons in the floor of the original valleys through which they flow, making this a country of high mountains and even deeper gorges, where the temperature can shift as much as thirty degrees in the course of a day's ascent or descent, and where travel was then a matter of precarious mountainside footpaths or shifting, boulder-studded streams. Not surprisingly, Wuting hsien to the east of the Chinsha and Hweili hsien to the west were among the centers of aboriginal population in the southwest, a distant backwoods where a mountaineer people could live under only the most superficial Han surveillance.

In the Kokiu region Ting had photographed and talked with several groups of aborigines. In Kunming, before setting off for the west, he commissioned some local metal workers to fashion him a caliper from a drawing in the

English Royal Society's *Guide to Travellers*. En route, where possible, he supplemented photographs and head measurements by interviews with individual aborigines and studies of their local records. As a peregrinating Han "official," Ting found grievance petitions pressed upon him by several tribal overlords; he received them with polite skepticism. His interest in aboriginal history did not prevent him from feeling that the future of the aborigines necessarily lay in education and assimilation into Chinese and modern society.

Ting's route led him past Wuting to Yuanmow, and on to cross the Chinsha River at the old courier outpost at the tip of its great horseshoe curve. From there he passed into Hweili hsien, a projection of Szechuan province surrounded on three sides by the river's enormous loop. After exploring the imposing Lung-chao and Lu-nan mountains, which encircle the hsien capital, Ting turned back toward the Chinsha, following tributaries which carve out steadily deepening canyons as they descend rapidly to the point of confluence. Crossing the river back into Yunnan once more, he detoured for twelve days to explore the uninhabited, unnamed, and impassable mountain range which lies between the P'u-tu River and the Hsiao River, two of the Chinsha's eastern tributaries. There Ting climbed Ku-niu peak (*Ku-niu chai*), which at 4,145 meters (13,595 feet) is the tallest mountain in Yunnan, and investigated the spectacular gorge of the Hsiao River nearby, China's deepest canyon, which cuts almost 4,000 feet deeper than the Grand Canyon of the Colorado.

In the course of his journey through Chinsha country Ting stopped to examine a number of small native copper mines, and at Ch'ing-kuang-shan in Hweili hsien he visited China's only nickel mine. Traditionally this mine had supplied Hankow artisans with a nickel-copper alloy known

47

as "Yunnan white brass," but recent competition from German suppliers had completely ruined the native nickel extractors. Ting found the original pits entirely flooded. In the derelict village an old man was able to outline the complex smelting procedures which had formerly been carried out on a site now little more than a shambles of potsherds.

After these exciting detours Ting finally proceeded to the places of his assignment, the copper mining center at Tungchuan and the coal fields in the neighborhood of Iwei. At Tungchuan, beside analyzing the deposits, he studied the history of the mines, which had operated under an imperial monopoly during most of the Ch'ing dynasty, producing copper for government mints. But now, in 1914, he was shocked by evidence of deplorable management. The old bureaucratic administration, hampered by centuries of accumulated customary procedures and struck hard by the disasters of the Taiping and Moslem rebellions of the mid-nineteenth century, had failed in its feeble efforts to modernize the mines. Recently it had virtually abdicated in favor of private speculators, who were taking advantage of the favorable market conditions prevailing after the revolution of 1911 to make enormous, quick profits, to the complete neglect of any program for long-term capitalization and development.[20]

From Tungchuan, Ting continued east, entering Kweichow at Weining hsien. Then he turned southward, passing through Iwei, Chu-ching, and Luliang, and so back to Kunming. He arrived home in Peking in February 1915.[21]

There Ting returned to his duties of training Chinese geologists. As a teacher his emphasis upon field work was especially striking, because at that time lack of first-hand observation was often a defect even of Western schools and because Chinese students were particularly ill-prepared for hard physical exercise. His students were taken out on weekly expeditions, and they were exhorted with the motto, "Always go on foot; reach the top of any mountain." Map-making and fossil-collecting were taught as routine parts of geological work. Moreover, Ting insisted that students achieve a thorough mastery of principles and foundations before attempting to produce original research from Chinese data.[22]

Although Ting enjoyed teaching, it was clear that the government's geology research organization could not continue indefinitely to expend its energies on pedagogy. In 1917, after Ts'ai Yüan-p'ei had arrived to take charge of a revitalized Peking University, it was arranged that instruction in geology be restored to the university and that Ting's organization devote itself solely to research under the name of "Geological Survey" (*Ti-chih t'iao-ch'a so*). Even after this transfer of duties Ting kept a watchful eye on the progress of the geology department at Pei-ta. In 1920 he gave the first group of university graduates in geology a short examination, and, discovering that they were by no means competent even in some of the fundamentals of rock classification, he decided that changes were necessary.

He found Ts'ai Yüan-p'ei most cooperative, and together they agreed upon two new appointments to the university: Li Ssu-kuang (J. S. Lee) and A. W. Grabau. Li, who at that time was completing his studies in England, became a noted petrologist; Grabau, already a distinguished paleontologist, was of German descent and had lost his post at Columbia University during a wave of war hysteria in America. Enrollments in geology began to rise and the department was on the way to maturity.[23] Grabau in particular became an important asset. He was given a joint appointment with the university and the Geological Survey, and his established international reputation added luster to both. Although he lectured in English to Chinese students who had never been abroad, and despite a disabling illness that prevented him from making field trips, his influence went deep. He won the respect and loyalty of his associates through his mastery of a critical subject and his uncomplaining performance of his academic duties in spite of Peking University's near collapse during the civil wars of the 1920s. Ting and others acted as legs for the Western scholar, whose papers on the fossils they brought in were among the ornaments of the *Paleontologica Sinica*.[24]

In 1924 Ting declared in a speech that he believed that in many respects training in geology at Peking University now could compare favorably with that of the universities and mining schools of the West. He thought that it surpassed most Western institutions outside the United States in its strong emphasis upon field work, while its gravest defect remained the lack of serious courses in the biological sciences at Pei-ta, which made it almost impossible for a student to understand the fundamental principles of historical geology. He was proud of the fact that the Chinese geologist needed a versatility and hardihood rarely demanded elsewhere. He had to know some surveying and be prepared to make his own topographical maps; moreover, in the field he had to put up with poor accommodations or none at all, lack of good food, and primitive methods of transport.[25]

By 1930 Ting was even more confident of the quality of geological studies at Peking University. Privately he told his friend T'ao Meng-ho that he thought the subject had reached the stage in China where it was no longer automatically necessary to favor the graduate of a foreign institution. From now on, he said, he expected the two groups to meet and compete as equals.[26]

In addition to decent training facilities a flourishing scientific discipline needs associations where specialists can gather to discuss and work in common, and it needs outlets for the publication of research. The Geological Survey itself was, of course, an important nucleus of research activity, employing many of the first Chinese geology graduates. One of its earlier tasks was to create a standard scientific terminology for the field. Ting and Wong Wen-hao opposed the nationalistic advocates of an entirely new "Chinese" terminology and instead advocated heavy reliance on expressions already developed in Japan. A brief dictionary of recommended translations from English, compiled by Tung Ch'ang with Ting's help, was soon widely followed. A long-term map-making project, on a scale of one million to one (the specifications of the International Geographical Association) was begun in 1924. In addition Ting rallied survey members, students, and friends to map wherever possible in the course of their travels and even to take bearings of latitude and longitude for general cartographical purposes. The results of mineral surveys began to be published in a series of "Geological Memoirs" (*Ti-chih ch'uan-pao*), of which the most important was the annual "Chinese Mining Bulletin" (*Chung-kuo k'uang-yeh chi-yao*), which first appeared in 1921.

Ting also took the lead in organizing two scholarly journals, the *Bulletin of the Geological Survey of China* (*Ti-chih hui-pao*), begun in 1919, and the *Paleontologica Sinica* (*Chung-kuo ku-sheng-wu chih*), which Ting served as editor for fifteen years, until his death. Because the first was published in English and Chinese and the second almost entirely in English, these journals quickly achieved international circulation.

Outside the survey there was room for an association of Chinese and the foreign scientists who were gravitating to Peking on a variety of projects. In those years The Rockefeller Foundation was sponsoring Peking's first modern medical school, where the anatomical department was headed by the anthropologist Davidson Black; the American Museum of Natural History had a mission in China; there was a French Jesuit scientific mission under Père Licent, a Swedish mission under Sven Hedin, a Geological Committee of the Russian Far East, and others. In 1922 Ting and his colleagues at the survey gathered resident Peking scientists into the international "Geological Society of China." Ting presided over an organizational meeting of twenty-six charter members in March. It was, Davidson Black said, "the first non-medical scientific association initiated and organized wholly by Chinese investigators for the study and advancement of one of the pure sciences." [27] The association grew rapidly and published a journal, *Bulletin of the Geological Society of China* (in English), and in the next years Ting served variously as board member, as editor, and for three terms, as president.

This organization provided the formal center for Sino-foreign cooperation, which was unusually free in geology and related fields. Ting had done much to make this possible, of course, by creating a Chinese geological organization whose members could approach Western counterparts with the self-confidence of colleagues rather than the hesitation of apprentices. Moreover, he consciously acted as an interpreter between the two groups, stressing the international nature of science and illustrating the possibilities of friendly relations in his own behavior. In a typical speech, delivered on the occasion of Davidson Black's death, he spoke of the trivality of concepts like nation and race compared with the pursuit of scientific truth, and he urged Westerners to forget their feelings of superiority and Chinese their hypersensitivity.[28] His cordial relationships with foreigners often amazed his Chinese friends. Grabau was an intimate friend, Teilhard de Chardin and Davidson Black were good ones.

Even so, scientific collaboration was not entirely unmarked by friction. Both Sven Hedin and Teilhard de Chardin met with criticism from other Europeans when they consented to work upon archaeological projects under Chinese, not foreign, direction. On the other hand, the Chinese authorities were suspicious of independent foreign scientific expeditions and especially objected to exportation from the country of valuable artifacts and specimens unearthed by outsiders. Ting had to urge the Chinese archaeologist Li Chi to overcome his hesitations about accepting a post with the Freer Art Gallery of Washington.[29] Constant awareness of the barriers to be transcended increased the self-consciousness of the little group of scientists of Peking, but it also sweetened a camaraderie precariously won in a world of hostile nationalisms. Teilhard de Chardin wrote feelingly of one particular meeting amidst the uncertainties of the Northern Expedition:

I have often told you of the cordiality of these gatherings in Peking. There were ten of us, all friends and almost all intimate friends; four Chinese: Ting, Wong, the director of the survey, Lee, a professor at the University, and King, an amateur of natural history; several Americans: Dr. Grabau, my great friend the paleontologist, Dr. Black of the American University of Medicine, and Granger, the paleontologist of the American expedition; two Swedes: Andersson and Sven Hedin; and myself. As usual, there was a more or less conscious feeling of achievement in meeting on the level of common humanity, man to man, transcending all national, racial, or even "confessional" barriers. But on that occasion, with the meeting a prelude to many partings, and in a Far Eastern atmosphere charged with passions and hatreds, a feeling of poignancy overlaid our talk. The table was strewn with apple blossom, a sign both of spring and of its storms; we each spoke in turn of friendship, of collaboration, of our hopes for small or great things.[30]

As the years passed, a series of directorships, editorships, and board memberships made Ting increasingly a man with the ability to get things done. He was an editor of the magazine *Science* (*K'o-hsüeh*); after 1923 a position on the board of the China Foundation enabled him to tap reserves of American money which could be relied upon to support scholarly projects when the government's treasury ran dry. The Foundation subsidized *Paleontologica Sinica* and gave the Geological Survey an annual subvention which almost doubled its financial resources. When the discovery of a humanoid tooth at Choukoutien in 1927 opened a brilliant future for the investigation of prehistoric man in China, it was Ting who arranged for the collaboration of the survey and the Peking Union Medical School, aided by Rockefeller Foundation money, in a special research project on Cenozoic man. He coaxed open the purses of native mining industrialists to finance a museum and library for the survey. He was a delegate to international congresses of geology and anthropology; he watched over Chinese students of geology abroad, whom he would advise as to specialization and commission to buy books; and he kept on the alert for foreign experts who might be willing to spend a period of research and teaching in China. The title of honorary director of the Geological Survey, which he retained until his death, represented not only etiquette but also Ting's lifelong informal supervision of the projects, personnel, and prospects of the survey, which by 1930 had spawned a flourishing tribe of dependent laboratories.[31]

As this record of his activities indicates, Ting maintained his position of scientific leadership and influence even though he left his post as director of the Geological Survey in 1921 to go to work as managing director of the Pei-p'iao coal mine in Jehol (Manchuria). He remained in private business for five years, until the end of 1925, and did not return to full-time academic research until the beginning of 1928. However, during those years he continued to live near enough Peking to make frequent visits there and to be in intimate touch with its scientific community.

Even by 1921, when he left the survey, Ting could take pride in a remarkable achievement: he had created the organizations and environment in which competent research in geology and other fields could be done in China by Chinese. He was sensitive enough to the conventional standards of international science to be apologetic about the fact that he had been a gadfly, a stimulator, and an organizer at the expense of any important personal contribution to pure scholarship. As of 1922, in addition to a paper on the stratigraphy of eastern Yunnan presented to an international congress in Belgium, Ting had published only five technical essays, mostly on the mineral resources of various districts of north China. But he remembered with a conscious sense of identification that Huxley had sacrificed his private research in order to act as the apostle of Darwinism in British society.[32] In fact, Ting's eclecticism could be considered beneficial to a pioneering scientific enterprise. His range of interests made him continually aware of the wider interrelations of his specialty and sensitive to discoveries profitable to related disciplines.

He carved out footholds not only for structural geology, but for mining engineering, physical anthropology, archaelogy, and cartography. Above all, his administrative talents and his interest in teaching were instrumental in making the survey and Peking University's geology department into viable institutions.

The most common danger to the whole first generation of Chinese foreign-educated specialists was the sense of futility and demoralization likely to overcome the trained man who was transported from the mature research atmosphere of his university or graduate school abroad to the actual conditions of his work back in China — isolated from colleagues, and deprived of adequate funds, literature, and equipment. Ting had been able to overcome these formidable obstacles for himself, and in so doing he had made them less daunting for others. Some external factors helped him. First, geology had a practical value, which generated a measure of official financial support. Also, in geology and the related historical natural sciences Chinese were provided with an ideal living laboratory — the land itself, whose exploration could easily and quite cheaply yield knowledge at the frontiers of research. Further, as an historical science, geology made use of empirical research methods which had some parallels with the concerns and methods of traditional scholarship. Essentially, historical questions were asked about origins and growth; results were formulated in ordinary, everyday language rather than in the abstract language of mathematics; and reasoning from individual data to generalizations and classifications usually took a roughly inductive form. But these factors need not detract from an accomplishment which owed much to Ting's personal combination of intellectual keenness and administrative flair.

In the years before 1949 the Geological Survey was a

source of great official pride to the Chinese, who spoke of it as a living illustration that given proper conditions of work they could match Western accomplishments in science. Later the Communists were inclined to dismiss the achievement of Republican geologists. Their more routine accusations — that Republican scientists had been self-centered, individualistic, poorly organized, and addicted to a bourgeois and classical ideal of theory over practice — bore the marks of party propaganda.[33] More seriously, the Communists denigrated geology as a secondary, historical, classificatory discipline, and they pointed out that a nation may not be considered to possess a mature scientific establishment until it has produced fundamental theoretical results in a basic discipline like physics or chemistry. However, the fact that pioneering Chinese researchers made their first contributions to pure science in one of the historical rather than one of the physical sciences may reflect upon the wealth and sophistication of the academic establishment in general, but it does not discredit the quality of work that actually was produced.

The Geological Survey had a legitimate position in the international learned world: its scholars were known; its journals were read; its research made a genuine contribution to knowledge of the natural history of the earth. Westerners called the survey the best scientific organization in Republican China, proof that lingering clichés about the inborn humanistic bias of "the Chinese mind" had decisively been rendered absurd.[34] In 1932 Dr. David White, principal geologist of the United States Geological Survey, wrote to Ting praising the Chinese Survey's "fine series of publications," and adding, "we marvel at what you are doing under circumstances that at best must be disheartening if not hopeless." [35] When he died in 1936, Ting was probably better known abroad than any other single Chinese scientist, mourned as "an enlightened pioneer who exerted great influence in promoting the development of science and its applications." [36]

For a dedicated scientist, Ting made an apparently curious decision in 1921 — to leave the full-time directorship of the survey to Wong Wen-hao in order himself to become the manager of the Pei-p'iao coal mine. He claimed at the time that he acted solely from personal motives—his need for a larger salary to finance the education of his younger brothers.[37] Ting always took his family responsibilities seriously, and in fact his younger brother Ting Wen-yüan did indeed soon afterward leave for several years of study in Switzerland and Germany. There is evidence, however, that he was also increasingly frustrated because of the hindrances to scientific work suffered under the reactionary militarists in charge of Peking's administration. It began to seem less and less likely that scientific research and education could enjoy healthy development in China unless they built on the foundation of a stable political order.[38] Concern for family and nation broke down the strict professionalism of Ting's early career as a scientist, and although he earned his living in the useful but less creative role of a private businessman, more and more of his energies were diverted to popular essay writing and attempts at political action.

While at the Geological Survey, Ting had, of course, taken a lively interest in practical mining developments. Mineral prospecting and the observation of existing mining conditions were the chief official purpose of most of the field trips he and his colleagues undertook. In *Miscellaneous Travels,* years later, he wrote from original notes detailed accounts of traditional mining methods as he had observed them in Shansi and Yunnan; more systematically, he gathered material on contemporary mining operations which he published in two pamphlets, *Materials for a History of Foreign Capitalized Mining* (1929) and *Mining in China During the Past Fifty Years* (1922).[39] Further, for five years, between 1921 and 1926, he was an editor of the Geological Survey's *Chinese Mining Bulletin*. Practical prospecting and planning took up much of his time. He made a detailed study of the coal fields of I hsien in south Shantung, and he drew up plans of extracting procedures for the Chung-hsing mine there, which later became one of the most prosperous native coal mines in China. In northern Chihli, where members of the survey had discovered substantial iron deposits, he served as one of the founding directors of the Lung-yen iron mine in 1920 and was active in establishing a blast furnace to service the mine at Shih-ching-shan near Peking.

The Pei-p'iao coal mine, located in Chaoyang hsien of Jehol province, was in 1921 a decayed official enterprise under the management of the government's Peking-Feng-tien railway bureau, which had originally set it up with capital of 500,000 yuan. After a Geological Survey report affirmed that the site was worth developing, plans for a reorganized mine were begun by Ting's friend and fellow provincial Liu Hou-sheng, a former associate of Chang Chien and a businessman who enjoyed the necessary official connections in the Ministry of Agriculture and Commerce and in the Ministry of Communications. For a Chinese mining company, the capitalization of the new firm was an ambitious 5,000,000 yuan, but it was typical of native ventures in falling far short of its financial goal: the actual sum subscribed was no more than 1,750,000 yuan. Although 40 percent of the reorganized mine's capital was supposed to be officially subscribed, it escaped the restrictions of official supervision and probably most of the benefits of official subsidy as well. In his own pamphlet, *Mining in China During the Past Fifty Years,* Ting categorized the Pei-p'iao mine as one of a group of twenty-three "important" native mines under private control.[40] When the company began operations in 1922, Liu Housheng was president, he himself was managing director, and the board of management was set up to be composed of seven merchants and four officials.[41]

Unfortunately little is known of the operations of the mine under Ting's management. Some of the technical problems involved may be inferred from Ting's remarks concerning the native and foreign coal mining enterprises which he observed in Shansi during the winter of 1913–1914. Throughout China the development of modern mining was handicapped not only by privileged foreign competition, but also by the fact that the vast majority of the more important mineral deposits had been worked for centuries by means of primitive surface extraction techniques and were ridden with traditional vested interests and outmoded customary law. The Republican administration's revised mining regulations of 1914 attempted to remedy matters by removing ownership of mineral rights from the control of surface landowners, but officials were hampered by limited administrative control. In coal mining, native extractors were able to obtain in usable form no more than 30 to 50 percent of the coal actually present in a given deposit, as opposed to 90 to 95 percent obtainable by up-to-date European methods; in addition they made a field unfit for further exploitation. After being hand mined, a field would be strewn with abandoned pits which gradually filled with water underground while their apertures became overgrown above, making it impossible for subsequent operators to avoid dangerous cave-ins and flooding. In Shansi the few existing Chinese semimodern mines represented only a marginal improvement over traditional ones in productivity and costs. At a typical small operation, the Cheng-ling Company, Ting observed in 1913 that mechanical aids consisted of no more than winch-driven cart, pump, and tiny railway, and he complained that the company employed no engineer, followed no extracting plan, possessed no map of the mine's interior, and refused to show him its production figures. In any case production could not be compared with that of the neighboring German-operated mine at Ching-hsing.[42]

By the late 1920s, shortly after Ting had resigned his directorship, the Pei-p'iao mine could be described as one of the more profitable smaller Chinese mines, one which enjoyed a degree of basic mechanization. From a hillside digging consisting of two inclined tunnels which yielded under 8,000 tons of coal in 1921, it had expanded into a mine working out of two vertical shafts about 700 and 100 feet deep respectively, capable of producing from 1,500 to 2,400 tons a day. The most important mechanical equipment consisted of an electric generating plant, which was used to operate small mechanical pumps, fans, cable railways and hoisters, and two mechanical sifters. Facilities at the mine included an infirmary, a repair shop, 315 five-room cabins for workers, and about 20 dwellings for management personnel; all in all, the mine employed about 3,700 laborers and a clerical and managerial staff of 110. Clearly, however, most of the actual coal cutting and probably a considerable amount of the hauling continued to be done by hand. For their labor workers received a minimum wage of thirty-five to fifty cents a day.[43]

This description of the mine dates from about 1930 and represents a considerable expansion beyond that which Ting could personally have overseen. However, production figures (in tons) for the ten-year period beginning in

1921 show that he laid the foundation for its development:[44]

1921	7,716	1926	153,462
1922	25,808	1927	286,087
1923	29,536	1928	367,009
1924	63,384	1929	406,427
1925	144,758	1930	509,872

Although the greatest volume of production was achieved after Ting left the mine at the end of 1925, he was effectively in charge during a period of formative growth.

Some of the serious obstacles to progress must have been political. Chaoyang hsien lay under the authority of the warlord Chang Tso-lin, governor of Manchuria, and Ting found it necessary to make frequent trips to Mukden to negotiate on the mine's behalf with authorities whose chief preoccupations beyond personal enrichment were the appeasement of Japan and military expansion south of the Great Wall. The mine was dependent upon the Peking-Mukden railway system, which was both its largest single customer and its means of transport to places like Mukden and Yinkow, where the bulk of its remaining output was either shipped or sold. Transport fees on the average doubled the total cost of production, and furthermore the railroad itself was unreliable in its operations. In 1921 the railway company had completed thirty-three miles of a Pei-p'-iao–Chaoyang branch line, but the building of the remainder of the track was repeatedly delayed by military disturbances in the autumn of 1922 and the spring and summer of 1923. By 1925 Chang Tso-lin's armed adventures had reduced the entire Peking-Mukden railroad to a state of disorganization which severely curtailed its functioning.[45] These facts suggest personal experiences animating Ting's often expressed distaste for the Manchurian warlord's brand of militarism.[46]

As a mining executive Ting tended to favor both private enterprise and some measure of foreign participation in the extractive industries, but he did so on purely pragmatic grounds: he thought that these methods did most to foster growth and technical efficiency under existing Chinese conditions. While denouncing past instances of foreign monopoly and criticizing extraterritoriality as an abiding obstacle to greater Sino-foreign cooperation, Ting nevertheless supported the Chinese government's policy of encouraging combined Sino-foreign mining companies which operated under Chinese law, and with at least a nominal 50 percent Chinese capitalization. In this regard Ting's nationalism gave way to his respect for expertise. Production statistics spoke for cooperatively managed mines: in 1922, of the twenty-two such comparatively successful ventures the Ministry of Agriculture and Commerce had figures on, the majority involved Sino-Japanese collaboration.

As for private enterprise in modern mining, Ting preferred it for equally practical reasons. State economic controls had for him neither the meaning given them by Western socialists preoccupied with the achievement of economic equality nor that attributed to them by capitalist champions of individualism. Such controls rather suggested centuries of Chinese imperial mining monopolies and the kind of inefficient bureaucratic manipulation and supervision of infant modern industries which had so notoriously failed under the "official supervision, merchant management" (*kuan tu shang pan*) system of the late Ch'ing. Government intervention in industry was not undesirable in principle; it had simply proved itself inept in fact. In the extractive industries the "official supervision, merchant management" system was condemned by the most unambiguous standard — its production record, which showed, according to Geological Survey statistics, that between 1872 and 1922 the vast majority of new mines which began operation under official control, or which allowed merchant participation only under official supervision, had not prospered. Of a total of one hundred such mines listed by the Ministry of Agriculture and Commerce in 1921, sixty-one either had ceased to function or had been turned over to merchant management, while seventeen more were known to be currently operating at a loss. The most successful category of official mine was that managed by private individuals, with output marketed entirely through official channels. Ting observed, however, that these mines profited by the exploitative expedient of denying adequate payment to workers.[47]

In the 1920s Ting thought political conditions in China were such as to make increased government involvement in economic affairs nothing but an obstacle to development: "In the present condition of our country . . . when private individuals reclaim salt land, 4 million *yuan* of capital is sufficient, but if the nation undertakes it, 8 million will not necessarily complete the work. Workmen repair a road at the rate of one *yuan* a foot [*chung*] and it may last for three years; roads repaired by soldiers cost two *yuan* a foot and still will not last for two. Where everywhere there exist scarcity of talent, immaturity of organization and evil practices among official cliques, can there be a good result?" An orderly caretaker government was the most that he asked for: "If finances are well-managed, property is protected and obstacles to communications are removed, there will be no problem about the national economy." [48]

His preoccupation with political barriers to economic

growth and his rather naive faith in technology made Ting seem for a time in the 1920s like an exponent of the strictest nineteenth-century capitalism. He maintained a simple confidence in the potential of the economy of China, a big country where ample natural resources were bound to exist, and which needed only to be liberated from traditional obstacles to economic growth. Until the Stalinist system of planning presented him with an alternative model for the industrialization of a nation and an illustration of its possible successes, he assumed that the prevailing European pattern of private enterprise, under the protection of stable governments, was the tested method for prosperous development.

However, in these years what passed for a Chinese government did not even command the necessary skills to create an environment favorable to economic growth. On the contrary, the conditions of his work at Pei-p'iao mines, like those which surrounded him as a government-employed teacher and researcher, indicated that an unstable political order hindered entrepreneurial development as it did educational reform. As long as these conditions prevailed, the economy would remain on the edge of bankruptcy and the hopes of men like himself would continue to be frustrated.

Chapter IV

The New Culture
and Chinese Tradition

During Ting's years at the Geological Survey, Peking was at the heart of China's developing new culture movement — the capital's intelligentsia were its natural leaders and the students its effervescent shock troops. In every field Peking University and the people associated with it took the lead, especially after 1917 when Ts'ai Yüan-p'ei reorganized the institution and brought to it some of the nation's finest foreign-trained younger scholars. Language reform, anti-Confucianism, liberal, socialist, and Marxist political ideas, enthusiasm for science — all diffused to the radical public from the circles around Peking professors like Hu Shih, Ch'en Tu-hsiu, Li Ta-chao, and T'ao Meng-ho, as well as from local journals like *New Youth* (*Hsin ch'ing-nien*), the *Weekly Critic* (*Mei-chou p'ing-lun*), and *New Tide* (*Hsin ch'ao*).

On May 4, 1919, the students and professors of Peking University gave China more than intellectual leadership. Sudden, shocking evidence of continued Western arrogance and Chinese abasement came in the form of news from France, where the delegates to the Versailles peace conference announced they had transferred to Japan the former German treaty privileges in Shantung. Students and faculty alike took to the streets in protest, sparking a wave of demonstrations which toppled a Chinese cabinet and showed how directly the new movement of intellectual emancipation was linked to political revolt.

Gradually yet inevitably Ting Wen-chiang was drawn away from his research and into this hectic world of cultural and political pacesetters. It was natural that a person of his background and position would eventually meet some of the dominant personalities among the intelligentsia. An acquaintance with T'ao Meng-ho led sometime in 1919 to one with Hu Shih, who soon became one of Ting's most intimate friends. The two men contrasted somewhat in temperament: whereas Hu was naturally optimistic, easygoing, and gifted with artistic sensibility, Ting was rationalistic, orderly, and somewhat puritanical, generally indifferent to beauty or the pleasures of good living. As the older by four years, Ting easily assumed a fatherly air toward his friend, earning thereby the affectionate yet slightly ironical nickname "big brother Ting." [1]

In that same year Ting also came to know Liang Ch'i-ch'ao, certainly the most renowned if no longer the most influential of the Western-oriented Chinese intellectuals, who stood as an elder statesman to the entire new culture movement. In 1919 Liang led an unofficial Chinese delegation to the Versailles conference, with the aim of fighting the unequal treaties and observing postwar conditions in Europe. Through a mutual friend, Hsü Hsin-liu, Ting was invited to join the party, on the theory that an expedition of this kind needed a scientist. Besides Ting and Hsü, other members of the group were Chiang Pai-li (Chiang Fang-chen), Liu Tzu-k'ai, Chang Chün-mai (Carsun Chang), and Yang Wei-hsin. In this way Ting found himself in Paris during May of 1919, personally involved in the diplomatic side of the great protest movement.[2]

Liang Ch'i-ch'ao, fifteen years older than Ting, had

lived through his formative political experiences during the 1898 reforms, when the intellectuals dreamed of no more than modernizing the imperial monarchy. A gulf separated the two generations: one formed intellectually by the old culture, having surrendered traditional beliefs with great difficulty and almost in solitude; the other more self-confident, able to study Western culture directly and to shed many traditions simultaneously with the dependencies of childhood itself. Ting approached Liang in a spirit of almost tender courtesy, acting in a small way as an interpreter of the West for the older man. In Europe he was Liang's translator in French and English and for a time gave him daily English lessons. Their differences showed in their reactions to the war — for the Great War caused Liang to despair of European civilization in its entirety, while Ting saw the catastrophe as a terrible yet avoidable consequence of European errors. Ting also advised Liang to give up the cruel life of politics in the future and to return to the scholarship more congenial to his emotional and generous temperament. Urging him to turn his abilities to applying modern research techniques to the study of Chinese history, Ting introduced his friend to English works on scientific and historical methodology.[3]

Meanwhile in Paris politics was inescapable, as the Chinese learned that the peace which was to bring national self-determination in Europe would involve no relaxation of imperialist privileges in Asia. Liang had done much to make Chinese participation in the war as an Ally palatable to the Chinese intelligentsia by arguing that China might in this way buy redress of national grievances at the peace table. Now he stood by helplessly as the victorious powers prepared to transfer Germany's defunct treaty rights in Shantung to Japan. All he could do was bombard the press at home with news of the Peking gov-

ernment's treacherous acquiescence in the Shantung transfer and urge the public to repudiate the treaty. His delegation lobbied fruitlessly among the assembled diplomats
concerning customs autonomy, extraterritoriality, the
Boxer indemnity, and foreign control of Chinese taxes and
railways. At home the news from France and the spectacle
of official submission to Allied and Japanese pressure finally produced the popular demonstrations of May Fourth
and the succeeding weeks, which gave the nation its first
experience of the political potential of the populist urban
mob.

If the year 1919 was a turning point in the political
consciousness of the Chinese people, it was something of a
watershed for Ting as well, bringing him the friendship of
public men and participation in the events of a grave national crisis. It quite naturally followed that he began his
own career as a popular essayist before the year had ended.
His first articles for the general public, the series "Eugenics and Genealogy," appeared in Liang Ch'i-ch'ao's journal
Kai-tsao (Reconstruction) in the latter part of 1919.[4]

Despite the excitements of the year Ting approached
the subject of China's intellectual renaissance as a self-
conscious rationalist, determined not to judge on the basis
of an unreflecting advocacy of any "Western" innovation,
but on presumably universal and scientific standards of
validity. Much in the radical doctrine of the age was indeed highly rhetorical, valuable as a concerted declaration
of emancipation on the part of Chinese youth, but deficient
either as analysis of Western institutions themselves or as
proposals for realistic social action inside China. "New"
was probably the most overworked adjective of the decade,
and the Western models for novelty were indistinctly, if
enthusiastically, perceived. Mr. Democracy often appeared
simply as the embodiment of a new way of life dedicated

to individual self-expression rather than family obligation, or of a utopian vision of social harmony and cooperation in a purified future society. Often science and rationalism were only slogans in an anticlerical war against superstition and authority; appeals to the rule of reason were made in the language of rhetoric, and invocations of a scientific world view were reminiscent of a philosophical syncretism familiar in the Chinese past.

Such flamboyance was little to the taste of a man conscious of the difference between himself as a practicing scientist and others who were merely propagandists for science. "Eugenics and Genealogy" included some gentle criticism of radical excesses and implied graver questions about some radical assumptions. It also presented for the first time Ting's characteristic view of the relation between science and the problem of social and political reform.

The human revolution implicit in the new culture movement had led Ting to reflect upon Darwinism's contribution to human engineering, the science of eugenics. Sir Francis Galton (1822–1911), Darwin's cousin, had invented this field of inquiry by not hesitating to extend the simple principles of natural selection unequivocally to civilized man. He thought that social customs influence breeding among human beings, bringing about racial consequences that scientists have a duty to study. In this way man might, by selective reproduction, learn to maximize the incidence of socially useful human types. Like most scientists of his generation, Galton assumed that basic qualities of character and intelligence, as well as those of physical appearance and health, are inherited. Moreover, he believed that the distinctive racial and psychological characteristics of the diverse geographical and cultural groups into which mankind is divided have come about as the result of natural selection operating in a particular physical and social setting. Galton became the pioneer in the use of statistical methods applied to human attributes, both physical and psychological, and eugenics so acquired a powerful research technique.[5] During Ting's years in England eugenics was a flourishing academic study, sustaining journals and professorships in England and America and supported by associated eugenics societies of interested scientists and laymen. Ting was well acquainted with the works of Galton and of his chief contemporary followers, Karl Pearson in England and C. B. Davenport in the United States.

"Eugenics and Genealogy" was first of all an elementary introduction to the biological theories needed to explain the basic machinery of reproduction, heredity, and evolution as then understood. Ting concluded that the science of eugenics rests upon seven proven scientific hypotheses:

1. Man is one kind of biological organism. Propositions true for biological organisms in general are true for man.
2. Biological organisms all form part of a single system.
3. Natural selection is the only method of evolution.
4. The effects of acquired characteristics can hardly compare with those of inherited characteristics.
5. Habits are not inherited.
6. Reproductive cells are transmitted perpetually.
7. A man's individual nature is inherited from countless ancestors; the extent of his inheritance is in direct ratio to the closeness of the ancestor.[6]

What implications does this science have for China? Immediately it deflates some of the most cherished aspirations of the new culture movement. Eugenics teaches that "society and education alone cannot be the method for fundamentally improving the human species."[7] True improvement will result when the superior men in society have many children and the inferior men have few.

If the road to social progress must take as its starting

point the fundamental biological condition of man, then the Chinese, in considering the adoption of new social patterns, must base their choices on the scientific evidence which points the way to biological progress. Ting thought the study of eugenics could lead to a proper understanding of the mechanisms of population growth and limitation. Population pressures in operation inside China since the Han dynasty were, he believed, one of the important causes of China's backwardness; permanent peace and progress would be possible only when the number of mouths to be fed had stabilized at a level in harmony with the productivity of the land. The solution, in China as in Europe, would have to be a scientific program of birth control.

In addition Ting suggested that eugenics could provide the basis for a rational reform of Chinese family life. "Talk of new thought, a new culture and our 'transitional age' is exceedingly fashionable, but to embrace an idea because it is new and reject one because it is old is the opposite of scientific." [8] For example, the merits of concubinage should be considered in relation to the average quality of childen born of these unions compared with those of legitimate wives; the merits of large families versus small families should be considered in the light of the comparative intelligence and vigor of siblings in large families as opposed to those in small ones, and with regard to the effect of repeated childbearing on the health of mothers; the most suitable age for marriage should be considered in relation to the age at which parents are most likely to produce vigorous children; marriages conforming to the strict intermarriage taboos of the Chinese should be compared for vigor of offspring with marriages of a closer degree of consanguinity in other cultures. Finally, Ting noted that Chinese family genealogies, in spite of certain obvious defects, provide a fund of data for eugenic research perhaps unequaled in the world, and he urged Chinese to keep these records carefully in the interest of science.[9]

The conclusion that social progress must ultimately depend upon the existence of fewer, but more upright, vigorous, and intelligent people, whatever its merit, is too harsh to make a very popular case for "scientific" solutions to human problems — nor is it basically scientific, since qualities chosen for selection in man, as in ornamental pigeons, will be freely arrived at by the personal standards of the breeder. The creators of China's "new culture" had to work with the human material at hand and if possible have faith in its value. In Ting's solution are curious echoes of that perennial Confucian reliance upon "superior men" to make the body politic healthy, with the biological creation of a superior stock substituted for the proper education of the existing one. Moreover, Ting compounded the difficulty of has goal with a daunting complexity of means: prolonged biological research was needed to teach men what cultural adaptations would be advisable. For reformers longing for an ideology which could be translated immediately into deeds rather than a methodology demanding patient inquiry, Ting's proposals must have seemed disagreeably pedantic.

Taken strictly as science, Ting's eugenic proposals left society suspended until the advance of science could provide the basis for a rational social order. They assumed in addition that such advances would supply conclusions about the goals of social change as well as the means of reaching them. Actually, certain social values, contrary to the ideals of new culture enthusiasts, were already implicit in Ting's very emphasis upon the concept of biological progress. In suggesting that eugenic standards be applied to the family, Ting was assuming that the institution of marriage existed for the benefit of the social group, through the bearing and training of socially desirable offspring, rather than as an association for the achievement of personal happiness. This was not far from the ideal at the basis of the old clan authority. Further, eugenics, with its doctrine of fundamental inherited human inequality, obviously unchangeable by social or educational nurture, gave little comfort to the proponents of popular government. Galton's analogy, repeated by Ting, was that of an athlete who can develop his muscular powers so far and no further. Rule by the best, in the sense of Plato's *Republic,* would necessarily be the recommendation of the convinced eugenicist; in "Eugenics and Genealogy" Ting indirectly indicated that he did not consider the old Chinese examination system inconsistent with this ideal, since it tended toward the selection of the more intelligent individuals to assume the responsibility of governing. In fact, his understanding of the science of eugenics suggested to Ting that the old social leadership of the Chinese gentry, through its monopoly of education and access to official career, had some foundation in a biologically natural right.[10]

In this way, the nature-nurture controversy had its social meaning in China as in Eurpoe, and Asian radicals were not slow to see the implications. By the late 1920s Chinese socialists, following Marx, were declaring themselves to be thoroughgoing environmental determinists, against those neo-traditionalists who praised Confucian marriage customs and the imperial examinations in the name of biological selection.[11]

In practice, of course, Ting did not base his personal conduct upon the immature concepts of scientific sociology. Although he remained convinced that scientific methods, properly understood, should act as a guide in solving

social problems, in his daily life he practiced a much
looser kind of rationalism. In choosing between Chinese
and Western customs, he said in *Reconstruction,* men
should follow the rule of reason. Circumstances allowed
him to carry out this axiom in his own life with unusual
freedom — or, to speak more accurately, they allowed him
free choice in matters over which "the rule of reason,"
however conscientiously invoked, seldom has sovereign
control. The private morality of a man of the "new culture," with all its tensions between tradition and innovation, emerges with unusual clarity.

Within three years of Ting's return from Europe his
father had died, making the son at age twenty-seven the
effective head of a large family of dependents, and in particular of five younger brothers, little more than children.
Three were half-brothers, born during his years abroad.
On the one hand this event spared him the distress, so
painful to many of his peers, of embracing a new outlook
at the cost of personal filial rebellion; on the other hand
it made him a family head himself at an early age, in
charge of the moral nurture of others.

Ting broke sharply with the past by consciously cul-
tivating in his young charges independence of judgment
as opposed to respect for authority. We were taught, a
younger brother recalled, to exercise our own independent
judgment on questions of politics, economic systems, ancestral observances, merit in art, marriage, and choice of
vocation. Moreover, Ting was sensitive enough to try to
be tactful in discussing such matters with his brothers, not
wishing to stifle their thoughts and opinions with the unconscious weight of his own family authority and worldly
success. Further, he taught them to disapprove of the
conventional style of Chinese politeness when it substituted for sincerity and forthrightness formal manners

unrelated to genuine feeling.[12] In sum, they were taught intellectual independence, freedom in expressing personal preferences, and frankness in personal relationships.

Yet certain questions of right and wrong were not considered matters of independent judgment. Industry, honesty, frugality, cheerfulness, and above all the sacrifice of the individual for the benefit of the group were virtues immune from the intellect's skeptical scrutiny. It did not occur to Ting that a life of selfishness and greed might be chosen prudentially by an individual as "rational" as himself. Social utility was his touchstone for determining moral conduct, and he assumed that group welfare would be accepted as the ultimate moral end by all rational human beings. He shared with the great Victorian agnostics whom he admired a conviction, sustained by the tradition into which he was born, of a fixed ethical order whose imperatives were created by the simple necessities of human cohabitation.

Darwinism merely provided him with an explanation of the origin of this order in terms of human survival characteristics. Ting was too good a scientist to believe in a teleology of evolution spontaneously tending to create higher forms of life. But man in society lived under special conditions requiring new kinds of adaptations. Since Ting shared Galton's belief in the dominant influence of inheritance, he was in a position to conjecture that ethical and religious beliefs are biologically conditioned instincts — the religious prompted by primitive man's fear of nature and the ethical by the characteristics needed to ensure family and group survival. These instincts, he pointed out, have even more primitive prototypes in the social habits of animals, ants, and bees. Now that mankind is learning through science to understand and to control nature, the religious instinct is losing its usefulness and might be expected to die out gradually: ethical beliefs continue to be valuable as men struggle to organize themselves into ever larger and more complex communities. As "biology" these reflections resemble Kropotkin or the speculative Huxley of *Evolution and Ethics*. Biologically they had been answered before, by Huxley himself, impaled upon his own contradictions in *Evolution and Ethics*: assuming that natural selection operates in man as in other species, the individuals who sacrifice themselves for the group diminish their reproductive vigor in comparison with others and lead to a waning of the characteristic in the group as a whole. But neither Ting nor Huxley seriously heeded this argument. Huxley ended his life believing that social necessity punished the wicked as surely as God's hand ever did, while Ting was equally convinced that the survival and prosperity of the social unit was a biological goal which in successful societies was reached because individuals were conditioned to sacrifice their individual desires. Variations in human conduct in this respect were determined by socioethical inherited characteristics, just as intellectual performance was conditioned by inherited reasoning ability.[13]

So in fact Ting taught his young charges to practice on the wider stage of nation and mankind many of those virtues which had traditionally been the ideal of Chinese family and village life. Theirs was an absolute obligation "to sacrifice the individual and immediate for the group and the long term." [14] Although he was totally indifferent to Confucian ceremonials and did not require conventional displays of filial piety from his younger brothers, Ting retained an underlying respect for the ancestor cult. It expressed in the form of superstitious ritual the bonds which hold families together and which demand the sacrifice of the individual for the good of the group. "I believe that the family clan system should be overthrown," he was capable of saying, "but I have no sympathy for those who evade family and clan responsibilities." [15]

His own sacrifice to the clan took the form of a lifelong financial burden, which was required to send four brothers and a nephew through school and university (one of them abroad) and to aid in the support of several indigent relatives. It was partly in order to meet this obligation, which cost him up to three thousand yuan a year, that he left his important scientific work at the Geological Survey in order to take the less congenial job of managing director of the Pei-p'iao mine. This was one occasion in his life when family duty to some extent overrode even his sense of social duty, and a certain guilt over his choice seems to have influenced his subsequent political activism.

As the individual must sacrifice himself to the group in the family, so Ting thought that society has the same right to expect sacrifices. Moreover, this is particularly the obligation of the upper classes. The poor are limited to a grim struggle for subsistence and are too dulled by poverty to achieve any understanding of important social issues. In return for his financial and intellectual advantages, the educated man owes his career to the public welfare. "Be ready to die tomorrow, but work as if you will live forever" was the motto which hung on the wall above Ting's desk.[16] His sense of financial rectitude was almost pedantic, making him, for example, always refuse to use free official passes on Chinese government railways. He considered it immoral for a man to have a standard of living higher than that required to enable him to work well. In his attitude toward public service Ting retained some Confucian attitudes, including belief in moral excellence and exemplary conduct as a basis for political authority. His code also reflected the contemporary public man's need to resist equally traditional political temptations to venality and the use of patronage.[17]

In sum, in his personal life as in his political inclinations this most Westernized of Chinese intellectuals was neither an individualist nor a democrat. The scientism and social Darwinism inculcated by his Western education left him in the mainstream of Chinese political thought, which has placed a specially selected bureaucratic elite at the center of the political process. He believed that biological principles and social utility confirmed much in the Confucian social ethic and so justified its broadening rather than its overthrow. Finally, he stood for paternalistic reform, under the leadership of the educated class, which in modern guise was expected to carry out the function of the ideal literati: administrator and moral guide. Social Darwinism at each step reinforced rather than undermined these attitudes, while his scientism gave him confidence that his lines of reasoning were both modern and correct. As a man of the May Fourth generation Ting displayed most prominently those attitudes which were destined to survive it.

The issues most dramatically associated with the new culture movement involved China's present and future: prescriptions for social and political reform drawn from various Western models and offered with an iconoclastic disregard for the traditions of China's past. As we have seen, Ting remained somewhat cautious in promoting change, and he was able to do so in the name of scientific rationalism rather than through an appeal to traditionalistic sentiment. There was yet another complementary intellectual preoccupation of the Westernizers of the May Fourth generation — the restructuring of Chinese tradition itself. Even as they called for a new Western-oriented education and culture, reformers sought to reinterpret Chinese history in ways which would integrate the parochial Chinese past into a worldwide context, and in so doing they would help make some of the values of that past seem compatible with the demands of the present. The development of the idea of universal history and the reevaluation of the Chinese historical record to fit in with this idea were projects which satisfied the intellectual need of Western-oriented thinkers to make use of the insights which were the inescapable conclusion of European historical experience and scholarship. Such projects also satisfied their emotional need to reconcile the conflicting claims to value of Chinese and Western cultures. Ting was an active promoter of the new historical perspective.

The idea of universal history advocated by Ting and his associates essentially was derived from the insights of biology: the human species was envisaged as a biological unit and the history of human societies was seen as proceeding along a path of progressive evolutionary development. Acting upon such assumptions, modern "scientific" historians called for the final repudiation of the ethical goals of traditional Chinese scholarship in favor of skepticism concerning everything save the facts of the past as scientific analysis could determine them. They demanded both a totally critical and dispassionate examination of traditional classical texts and also the extension of historical investigation to extracanonical sources in folklore, literature, anthropology, and archaeology. Increasingly it was assumed that the application of these techniques would reveal that Chinese society had evolved in successive stages paralleling the stages of evolution observable in other parts of the world.[18]

As a Darwinian and a scholar aware of recent work in anthropology and the history of early man, Ting was particularly attuned to these global perspectives. He contributed to their dissemination in China by helping Liang Ch'i-ch'ao prepare a translation of H. G. Wells's *Outline of History*, which dazzled thousands of Chinese readers with its evolutionary vision of the unity of mankind.[19] In his own interpretation of the course of Chinese history, Ting was especially concerned to discredit traditional historiography and to emphasize instead the "slow but continuous growth" of Chinese institutions and the extent of outside influence upon them. The old orthodoxy, with its belief in a Confucian golden age of the past, curiously echoed in Western theses about the stagnant and immemorial East, had, he believed, prevented men from properly understanding the true evolution of China's monarchy and bureaucracy, its changing systems of land tenure, and its gradual colonization and assimilation of the aboriginal peoples of the East Asian mainland.[20] Moreover, from neolithic times onward the Chinese in fact had been receptive to foreign innovation from every worthy source that had presented itself, and they had adopted things as important to their culture as bronze and iron technology, rice culture, the Buddhist religion, and Arab and Jesuit mathematics. Nor had this intercultural communication proceeded only in one direction. The truth was that China, which superficially had seemed to be a unique, self-contained civilization, had really been open to the mainstream of global history all along, and the present period of East-West interaction was neither new nor threatening. "Let us remember that culture is the common product of mankind," Ting advised, and he meant this maxim to calm the uneasiness of Chinese forced into an emergency twentieth-century renovation of their particular culture: "Modern industrial civilization, true enough, was developed in Europe, but it could never have come into existence without the cultural capital of the preceding ages which was the common inheritance of the Eurasian continent, and to which China contributed her noble share . . . Let us . . . face our new environment without false pride, but with calm confidence, for to adapt ourselves to the conditions created by the industrial civilization which is conquering the world is not only our clear duty but also our indisputable right." [21]

Historical reevaluations like these helped to make Chinese history interpretable in relation to newly perceived worldwide patterns. Moreover, such insights supported the thesis that there existed a cultural parallelism between China and the West. The idea of evolutionary progress applied to Chinese history implicitly defended the creativity and adaptability of the Chinese people and their institutions, and it made current worldwide inequities in wealth and power seem to be matters of merely relative importance. In their search for an historical formulation which would reveal the pattern that progress had taken on Chinese soil, scholars assumed that Chinese experience ought to be shown to follow a path similar to the one the West had followed. It seemed particularly important to establish a parallel which would locate in Chinese history periods of intellectual vigor which were the stimulus to cultural and social innovation. In 1902 Liang Ch'i-ch'ao had been the first to propose a concrete equivalence between a certain period of Chinese history and a Western one when he identified the age of revival of "pure" classical learning in the early Ch'ing dynasty as China's indigenous Renaissance — a time in which a programmatic concern to rescue the true meaning of classical texts from corruption had in fact stimulated vigorous intellectual innovation.[22] Hu Shih, studying the long-submerged non-Confucian schools of the Chou dynasty, thought he had discovered a many-faceted "classical" age, in which logical and political thought revealed parallels with ancient Greece and anticipations of modern pragmatism, but which was later stifled by Confucian orthodoxy and Buddhist "medieval" obscurantism.[23]

The idea of the Chinese Renaissance had particular appeal, and since it referred to times closer to the present, it almost immediately came to be associated with the new culture movement itself. When students and intellectuals wrote essays praising the Chinese Renaissance, they were often thinking primarily of their own time — of the literary reform, the campaigns against Confucius and the monarchy, the introduction of Western science and technology, which were making the present an age of dramatic release from ancient routines. But they also wanted subtly to expand the concept, suggesting hopefully that the present troubled emancipation was part of a long-term development which had intimate associations with the purely Chinese past. Hu placed the first stirrings of the Chinese Renaissance in the time of Chu Hsi; Liang asserted that it had begun in the seventeenth century with the empiricist "investigation of evidence" (k'ao chü) school of Confucian textual scholarship, which by its essentially scientific methods of thought had prepared the Chinese for the later blossoming of a twentieth-century High Renaissance devoted to the adoption of Western science itself.[24]

Ting accepted the idea of a "Chinese Renaissance." It involved an account of the Ming and Ch'ing periods which claimed equivalence between Chinese and European intellectual movements while at the same time resting these claims upon a rather traditional historiographical substructure. Ch'ing historians were no different from the writers of earlier orthodox Chinese histories in bolstering the legitimacy of the current reigning house by suggesting that the preceding dynasty so rudely supplanted owed its decline to excesses of misrule: the intellectual, and hence moral, errors of an earlier age supplied in Confucian eyes quite an adequate explanation for the cause of political decay. Moreover, in the seventeenth century genuine Han patriots claimed that decadence of the literati had been responsible for the calamities which befell Ming China at the hands of the Manchu invaders, and the way was made clear for a generally accepted orthodox historiography concerning Ming decline. This in turn led to praise of the "investigation of evidence" movement introduced by Ch'ing scholars who explicitly repudiated the excesses of speculative neo-Confucianism which were believed to have undermined the health of the Ming state.

Ting, then, was inclined to describe late imperial Chinese history in the following way: after the waning of the liberal and artistic civilization of the T'ang dynasty, a reaction set in under the Sung, when scholars in the name of neo-Confucian philosophy (*li hsüeh*) poisoned intellectual life with a stultifying formalism. Ostensibly devoted to goals of metaphysical enlightenment adapted from the ideals of Buddhism, neo-Confucianism actually was an anti-intellectual "scholasticism without method, a religion without faith." Nevertheless its version of Confucianism succeeded in becoming a state dogma which was imposed upon succeeding generations through the stereotyped literary formulas learned by candidates for state examinations. However, "Towards the beginning of the seventeenth century . . . intelligent men began to become dissatisfied with the intellectual life of their time . . . Political dissatisfaction stimulated intellectual activity, and the spirit of inquiry, awakening from its prolonged sleep, was once more walking abroad. This great movement may be termed the Chinese Renaissance, which began by textual criticism of the classics and ended in the introduction of western science and philosophy." [25]

Now in fact the quarrel between the Ch'ing empiricists and the Ming metaphysicians took place well within the framework of accepted neo-Confucian philosophical canons, all of which went back to Chu Hsi. It revolved around the basic neo-Confucian dilemma of whether "principles" (li) lie in "things" or in the "mind." The doctrine that "principles lie in things" was ontological, and not about nature; it did inspire Ch'ing scholars to undertake scholarly analyses of ancient texts and inscriptions, but they were motivated by a quest for the pure meaning of classical canons, not by scientific curiosity about the natural world.

However, Ting, like Hu Shih and Liang Ch'i-ch'ao, claimed that the Ch'ing scholars had pioneered the scientific method in China. They accepted the idea that the requirements for scientific thought had been satisfied when an investigator focused upon diverse instances of empirical factual data and applied experimental procedures involving induction to make generalizations concerning such data. For Hu Shih the classic example of the method of "investigation of evidence" was the collation of ancient poetry texts performed by Ku Yen-wu to establish through their rhyme schemes guides to the archaic pronunciation of Chinese words. Hu also made enthusiastic claims for the scientific character of the writings of Tai Chen and Ts'ui Shu. He was aware that the conclusions of these scholars were often marred by the fact that they were insufficiently critical of the data they accepted as "evidence": obviously a forged or spurious text or an undated artifact could not be put to any reliable use in an argument which did not take their status into account. Ting also knew that the ambition which inspired such research was a quite un-scientific desire to fathom the true principles of classical wisdom. Nevertheless, he was willing to claim that their method was highly systematic and essentially scientific.[26]

Ting made his explicit claims for the scientific pedigree of Ch'ing scholarship through efforts to popularize the works of two more marginal figures, Sung Ying-hsing (c. 1600–1650) and Hsü Hsia-k'o (Hsü Hung-tsu, 1586–1641). Both men lived approximately in the time of transition between the Ming and Ch'ing dynasties, which was presumed to have coincided with the birth of a Renaissance spirit in China. More important, they broke with the dominant text-centered traditions of research in favor of the direct investigation of things — natural phenomena and man's technology. Hu Shih claimed that scientific research enjoyed a period of indigenous development in China largely by ignoring the motivations and presuppositions of the textual scholars and focusing upon what he conceived to be their inductive and experimental techniques for handling "evidence." Ting chose the more sophisticated examples of individuals who were suspicious of literary tradition on subjects of natural history, where truth could better be decided by firsthand observation. Moreover, by taking observed phenomena rather than written statements as their "evidence," Ting's protoscientists were less likely to be misled by reverence for Confucian canons or the concealed inaccuracies of texts. But Ting's thesis was hampered by the fact, too bald to be evaded, that both Sung Ying-hsing and Hsü Hsia-k'o had been men of no important intellectual influence, considered eccentrics by their contemporaries and descendants alike. Ting's contribution was to help rescue their writings from complete obscurity and to make their true accomplishments known to the first generation of Chinese really capable of appreciating them. In practice he proved chiefly that the appearance of a man with the interests of a scientific naturalist had been an intellectual possibility in traditional China even if it had not been a pattern.

A geographer and explorer, Hsü Hsia-k'o had fascinated Ting ever since he learned on his first trip through Yunnan in 1911 of Hsü's journeys through the southwest in the seventeenth century. A memorial tablet to Hsü written by a friend made it clear that the aim of his many travels was the accumulation of geographical knowledge: "Hsia-k'o was not happy with the sayings of the schools of magic and divination . . . He said, 'When the ancients wished to know of the heavens and earth, they often relied upon inherited and distorted interpretations. Even the records of the two rivers and the three mountain ranges are limited to China proper and do not survey distant places.' " 27

After four years of arduous exploration in Kweichow, Yunnan, and the region around the Chinsha River, Hsü painstakingly recorded his results in a series of travel diaries, originally published under the titles *Investigation of the Source of the Yangtze* (*Chiang yüan k'ao*) and *Investigation of the P'an River* (*P'an chiang k'ao*). Although these diaries were known to the conventional scholarly world chiefly for excerpted snippets of elegant landscape description, Ting felt that in their entirety they read more like a naturalist's field notes. He praised them for a veracity of observation which enabled him as a reader to identify geological features from the language alone and for a systematic terminology which assigned specific topographical meanings to descriptive nouns used by traditional writers in quite indiscriminate fashion. More important still, the diaries recorded discoveries of fundamental geographical importance: that the P'an River is the main source of the West River, that the Mekong and the Salween are in fact separate rivers, and that the Chinsha, rather than the then conventionally designated Min River, is the true source of the Yangtze. To Ting, the chain of reasoning which Hsü used to defend his hypothesis about the Yangtze represented "an utterly scientific piece of writing:"

The Yangtze and the Yellow Rivers flow through the south and the north respectively to reach the ocean. My district is located just where the Yangtze rushes into the sea . . . Natives of the place gaze across the vast waters and sail over them, knowing their greatness but not their length. In tracing the river's source they simply believe it arises in the Min Mountains [in Szechuan]. When I first examined the records I saw that the Yellow River enters China from Chishih . . . and that everyone says its source lies north of the Kunlun Mountains [in Tibet] . . . Could it be that the Yangtze has a shorter course than the Yellow River? Could it be that the Yellow River is twice as large as the Yangtze? When I traveled across the Huai and the Han Rivers and came to look upon the Yellow River, it was not a third as wide as the waters of the Yangtze. Could it be that the watershed drained by a river as large as the Yangtze was less than that drained by the Yellow River? After I had traveled through Shensi on the north, the five mountain ranges of the south, and over the Shihmen Mountains and the Chinsha River in the west, I then realized that the Yellow River drains five Chinese provinces . . . and the Yangtze drains eleven . . . I judged that the Min River from Chengtu to Hsü is not a thousand *li* long, while the Chinsha stretches over two thousand *li* from the Li River through Yunnan and the Wumeng Mountains to Hsü. Is it possible to reject the longer river and to honor the shorter one as the source of the Yangtze? [28]

A reader could think Hsü was making a semantical point about river names: a river's longest tributary conventionally is designated as its source. However, the passage is easily open to the interpretation which Ting gave it: Hsü was referring to an hypothesis he formed about the Yangtze and Yellow rivers after observing their mouths, and which he later verified experimentally in the course of his explorations. Hsü Hsia-k'o was a true scientific geographer, Ting said. "The *Zeitgeist* of the Renaissance was in him," he concluded, "when he braved physical discomfort to the point of starvation for the sake of pure knowledge and intellectual satisfaction." [29]

If Hsü Hsia-k'o was Ting's precursor as a student of geography and natural history, Sung Ying-hsing prefigured his interest in mining technology. Once again Ting learned of the earlier scholar in the course of his own travels in Yunnan, this time from a provincial gazette containing excerpts from Sung's descriptions of local copper metallurgy. The selections came from a long work published in 1638 on Sung and Ming industries, the *T'ien kung k'ai wu;* by the nineteenth century this book had completely disappeared in China. After several years of searching, Ting located an inferior Japanese reprint of the book dated 1771, and for a time one of the many projects on his desk was that of preparing a restored and annotated text for republication. He was anticipated, however, by the bibliophile T'ao Hsiang, who in 1929 published a new edition based upon the Japanese text together with excerpts from several Chinese encyclopedias, to which Ting contributed an appendix and a short biography of the author.[30]

The *T'ien kung k'ai wu* consists of eighteen sections which describe, as Sung observed them, indigenous Chinese techniques employed in crop raising, weaving, well drilling, irrigation engineering, metallurgy, and salt manufacture, as well as methods for making, among other things, pottery, jewelry, paper, wine, and firearms. Ting said of it that it was written in a spirit of opposition to the petty practices of the Ming Confucian scholars, and he thought the work disclosed "an underlying harmony with the modern scientific method." [31] Ting praised Sung for insisting upon direct observation of the activities and artifacts of daily life — things which scholarly convention deemed to be vulgar — and for remaining skeptical about the superstitions and folklore which played such a large part in popular accounts of natural resources and manufacturing. Because Sung understood the importance of empirical mathematical calculations in the profitable management of crops and the successful manufacture of metalware and porcelain, Ting even attributed to him a highly modern "statistical point of view." He also thought it was important to add an appreciation of Sung's moral integrity as a scholar: without regard for popular fashion, he had pursued his researches with perseverance and disinterestedness through years of political chaos and uncertainty not unlike those which harassed contemporary scholars in the 1920s.

It is possible to exaggerate the weightiness of these attempts by Ting and Hu Shih to give to science a native Chinese pedigree. They were certainly more guarded in their claims than the Liang Ch'i-ch'ao of *Intellectual Trends in the Ch'ing Period,* for whom the scientific spirit often appeared to involve no more than fidelity to facts and freedom from dogmatism. As we have seen, when he was forced to a close scrutiny of Ch'ing learning, Hu was ready to acknowledge in its creators inadequate methods and antiquated motives. As a scientist Ting felt most affinity with two individuals whose naturalistic interests placed them outside the mainstream of traditional scholarship. When he did refer to the "rigorously scientific methods" of Ch'ing luminaries like Ku Yen-wu, Yen

Jo-chü, and Tai Chen, he explained their use of such techniques partly on the grounds that the scholars had been under the influence of Jesuit mathematics, a claim which is still difficult for Western scholars to judge.[32] Ting and Hu were not entirely insensitive to the charge of "forced interpretation" hanging about their description of an indigenous "Chinese Renaissance." In the last analysis their claims for native Chinese science rested upon the hypothesis that scientific thought essentially is characterized by the use of induction.

In the writings of Ch'ing scholars there did occur instances of inductive reasoning leading to the classification of data and the derivation of empirical generalizations from them. Hsü Hsia-k'o's geographical analysis of China's river systems, cited by Ting, provides an excellent example. In the case of textual scholars, their application of inductive procedures might seem less ambiguous if considered from a purely internal point of view: by focusing upon the validity of a reasoning process alone, given the data already accepted as "evidence," without questioning the wisdom of using such pieces of "evidence" in the first place.[33] Ting was always ready to accept such restricted cases of inductive reasoning as satisfying the requirements for the scientific method.[34]

In this way European thought supplied Ting with entirely plausible reasons for stating that Ch'ing scholars had been precursors of science in the Western sense of the word. European thinkers trained Ting to believe that scientific thought is fundamentally a question of methodology and that this methodology consists of empirical, basically inductive procedures. Anglo-Saxon philosophers of science like Huxley and Karl Pearson, whom he accepted as intellectual models, were responsible for this account of science, and it was not inconsistent with Ting's own practice as a geologist. As the touchstone for science, he was encouraged to focus upon a highly abstract and logically imperfect philosophical account of a way of thinking, and not upon any concrete research activity seen in its entirety. Ting, faithfully applying this theory to the alien world of Chinese scholarship, was trapped by the limitations of his model.[35]

If his Chinese pride made Ting want to find native precedents for science, Western example made it easier for him to think he had done so. He praised Ch'ing scholarly techniques according to what he took to be the most modern standards, and in the eyes of other Chinese, not the least of whom must have been his friends Hu Shih and Liang Ch'i-ch'ao, his prestige as a scientist gave authority to that praise. The search for precedents for science in classical learning, which was such an important part of the current academic drive to "reorganize the national heritage," gained weight from the imprimatur of a prominent scientist in person.

However, in talking about the "Chinese Renaissance" Ting spoke most feelingly not about methodology but about the emergence in Confucian China of an intellectual spirit of inquiry which represented for him a timeless, universal ideal. He described scholars formed by China's traditional culture as having participated in the universal search for truth, which was contemporary scientific man's most lofty enterprise. Sung and Hsü were presented as models bridging the gap between the aspirations of Ting's generation and those of his ancestors, demonstrating that a modern Chinese intellectual could still be linked in spirit with scholars who as Chinese were his cultural forebears. There was a literal appropriateness in holding recent scientific intellectual ideals up as a standard by means of writing essays about the lives of seventeenth-century Chi-

nese gentlemen. A good orthodox Chinese biographical
essay was meant to provide readers with didactic examples
of conduct to be emulated. In his praises of Sung and
Hsü, Ting thought it important to link their intellectual
accomplishments — identified with the universal modern
standards of science — with their practice of personal virtues identified with the universal Confucian standards of
self-sacrifice and disinterestedness. In so doing he presented
in familiar literati robes exemplars of universal values
shared by both traditional China and the modern West.
However, where the generation led by K'ang Yu-wei which
had immediately preceded him faced the problem of uni-
versalizing Chinese values by depicting a utopian Confucianism as the global pattern for the future, Ting was
sufficiently distant from Confucian intellectual ideals to
feel required to reverse this process. He took the scientific
values which he assumed to be universal and by means of
the idea of the Chinese Renaissance projected them upon
the Chinese past.

Chapter V
Science and Metaphysics

Ting Wen-chiang's contributions to the early literature of the May Fourth period were modest, and they identified him with the more moderate academic intellectuals. For these men, education and study were the fundamental tasks of the new culture movement, in the hope that foreign ideas might receive balanced consideration and the assimilation of these ideas bring about the natural evolution of a rejuvenated Chinese society. However, events after 1919 put the entire movement under centrifugal stresses, precipitating sharper conflicts within the intelligentsia itself and forcing harsher choices upon its factions.

The May Fourth demonstrations themselves stood in striking contrast to the bookish preoccupations of scholars and suggested hitherto undreamed of possibilities for direct political action. In their wake, political militancy grew, and with it a quickening of interest in revolutionary ideologies. To enthusiasts for democratic, anarchist, or socialist utopias, Ting's advocacy of scientific sociology did not seem very satisfactory. Similarly, radicals like Wu Chih-hui ridiculed his efforts, along with those of Hu Shih, to link certain new cultural ideals with themes in the Chinese intellectual tradition, calling them attempts to whitewash Confucianism, unworthy of an emancipated individual.[1] Although Ting was not unaffected by the new mood of activism,[2] he was increasingly estranged from the rebellious iconoclasts of the far left.

On the other hand, May Fourth zealotry at home and the conditions of postwar Europe abroad were provoking a more serious return to conservatism among segments of the previously pro-Western intelligentsia. Suddenly the entire value of Western scientific and industrial civilization was being questioned again, this time not by Confucian diehards, but by foreign graduates themselves, men renowned as students of Western thought. As discussion grew more intense, Ting found his own views diverging from those of the right as well as the left. Finally he felt it necessary to differ from close associates and to battle openly in defense of the scientific outlook which to him was Europe's enduring contribution to world civilization. In 1923 debate crystallized in the famous polemic on "science and metaphysics," which flooded the journals with dozens of articles and brought up a wide range of philosophical, scientific, and social topics. As the foremost champion of science in this controversy, Ting made his real public reputation as a leader of the new culture movement.

The end of World War I marked a certain dividing point in the history of intellectual interchange between China and the West. It coincided with the later middle age of the first generation of Chinese reformers, who had crusaded for foreign learning to revitalize the imperial monarchy. In 1918 the actual success of some of their efforts was beginning to present them with unwelcome consequences never visualized in the pages of Spencer, Rousseau, or Mill. Moreover, the strain of attempting a personal synthesis between two cultures was beginning to tell upon men whose intellectual convictions had become divorced from the Confucian folkways which still provided their most intimate moral supports. Finally, the monolithic West, which had first appeared to China to be as united in its intellectual heritage as in its material power and technology, was fragmenting before their eyes. The more they studied Europe and America, the more they saw of schism and dissension and of uncertainty about the future almost on a par with China's own misgivings.

In China, modern reforms had indeed been carried out, but they led to unforeseen dislocations. The constitutional republic, the most ambitious innovation from the West, had also been the most crushing failure. There were also indications that the industrial complexes beginning to spring up in the treaty ports brought economic imbalance, not increasing abundance, and instead of opportunity for their workers, a life scarcely less grim and penurious than that on the bankrupt farms of the interior. Young people fortunate enough to receive the Western education being provided in a few privileged schools were showing signs that foreign learning produced discontent as well as new skills, inspiring an unruly impatience with the social sanctities of the older generation. Most dramatically, the nation's great experiment in international relations as a partner with the West — participation in World War I — had ended with the shocking disillusionment of Versailles.

Chinese intellectuals were also beginning to gain a more intimate knowledge of Europe. Their ideas had been formed by reading the classic authors of the European Enlightenment and those of their successors among the nineteenth-century scientific rationalists — Bentham, Mill, Comte, Spencer, Huxley. From half a globe away, the power and material abundance of Europe, its orderly habits of constitutional government, and the technical efficiency of its business and commerce had all seemed to be signs of a stable and unified civilization. But by 1920, after a generation of intensive study of the West, Chinese could not help becoming aware that many Western institutions were the focus of bitter controversies at home and that the ideas which they had first assumed were the foundation of Western power often were in conflict with one another. Nationalism seemed an incentive to war and colonialism; democracy, the beginning of a road that led to class conflict; freedom, a value that was fighting for its life before a trend toward protectionism in trade, authoritarianism in politics, and socialism in the field of human welfare.

Chinese heard European Marxists on the left condemn capitalist forms of industrial production as exploitation, which made the basic relations between men those of greed and dependence and enriched the privileged without really benefiting the masses. They heard that constitutional governments were the tools of class manipulation and that the economic logic of capitalist exploitation pitted strong against weak all over the world, leading in international affairs to imperialism and war. From the European right they caught echoes of a parallel dissatisfaction with contemporary industrial society, this time expressing itself in the form of protests against the dehumanizing influences of technology and through an intellectual rebellion against rationalism which took many complex guises. They saw embattled humanists retreat to existential and idealist philosophies, artists create a religion of aesthetic values, while authoritarian political prophets preached an anti-intellectual cult of freedom and power. Finally, the destructive instruments of technology and the uncurbed follies of men had bred a palpable monster, the Great War, which seemed in East and West alike to stand as an indictment of an entire civilization.

Liang Ch'i-ch'ao's trip to Europe in 1919 provided the occasion for the first important statement of the new skepticism concerning the direction of European civilization. Although Liang had been the most famous Chinese interpreter of the West for over twenty years, this was his only visit to Europe itself, where he found men in a post-war mood of exhaustion and futility and gripped by the passions of Versailles. The European philosophers he talked to, who included Henri Bergson, Rudolf Eucken, and Emile Boutroux, offered few words of reassurance or confidence. The essay which Liang published on his return, "Reflections on a Trip to Europe," showed how disturbing the experience had been.

Science and industrial revolution have transformed man's inner and outer life, he wrote. Before the development of science, European man had unshakable authority for his moral life in the precepts of Christianity and of Greek philosophy. But the triumph of science has done away with man's belief in heaven or in his own soul. According to scientific psychology man's mind and spirit are only phenomena of "matter in motion," and the great principles of the universe are arrived at solely by means of the scientific method of experiment rather than by the philosophical method of abstruse meditation. The result has been to create a "purely material, purely mechanistic view of man" (*jen-sheng kuan*) which subordinates man's life to the "necessary laws" of matter in motion. Ancient Chinese fatalism (*yün-ming ch'ien ting*) declared that destiny is fixed by the decree of the gods or the lot of numbers; Westerners have merely created a disguised new fatalism devoted to the proposition that destiny is entirely controlled by scientific law. In this way freedom of the will is denied, but without free will (*tzu-yu i-chih*) there can be no responsibility for good or bad. It seemed to Liang that in place of the old moralities based upon spiritual authority Western science offered a naturalistic morality of Darwinism, which worships power and money, pits the strong against the weak, and encourages imperialism and war. At the mercy of machines, guided by no principles save fear and greed, men become isolated, skeptical, enervated, and corrupt, and their lives are without enjoyment or value.[3] Liang's sense of poetry and his profound sincerity lent power to his bleak portrait of urban industrial man and urgency to his conclusion that the Chinese were headed for disaster if they followed Europeans in believing in the omnipotence of science.

The theme of European inadequacy was quickly taken up in a more systematic form by Liang Sou-ming, who in 1921 published an ambitious comparative study of *Eastern and Western Civilizations and Their Philosophies*.[4] After formulating an abstract characterization of the ideal essences of Chinese, European, and Indian cultures, he offered a cyclical view of three stages in world historical evolution: mankind would move from a "Western" civilization of struggle for gratification, to a "Chinese" civilization of moderation and contentment, ending in an "Indian" civilization of self-abnegation and mystical contemplation. Western materialism had found a revived challenger in the "spiritual East."

In debating the subject of Eastern versus Western civilization, the common Chinese assumption was that the characteristic social arrangements of each civilization must have their roots in a set of metaphysically prior philosophical and ethical principles. In trying to discuss issues of European and Chinese philosophy they hoped to expose the underlying forces responsible for two world systems and to reveal the primary axioms upon which a future Chinese society would have to be based.

Formally the issue was joined in a speech given before the students of Tsinghua University in Peking on February 14, 1923. The speaker was Chang Chün-mai (Chang

99

Chia-sen, Carsun Chang), the philosopher and close associate of Liang Ch'i-ch'ao. Ting knew Chang personally, since all three men had been together on the journey in Europe in 1919; Chang in fact had only recently returned from that trip, which he had extended in order to study philosophy at Jena under the idealist thinker Rudolf Eucken. The *Tsinghua Weekly* (*Tsing-hua chou-k'an*) quickly printed Chang's lecture, and a reply by Ting Wen-chiang in *Endeavor* (*Nu-li chou-pao*) prompted a flood of argumentation from several dozen writers, all concerned with the fundamental meaning of Western science and its implications for China's modern culture.

The question that Chang Chün-mai posed for the Chinese public was: "Can science govern a view of life?" His negative reply was summarized in his famous formulation that a view of life is "subjective, intuitive, synthetic, freely willed, and unique to the individual," in contrast to science, which is "objective, determined by the logical method, analytical, and governed by the laws of cause and effect and by uniformity in nature." [5] Chang's basic contention was that certain questions cannot be answered by an appeal to scientific reasoning, specifically those which have to do with the essentially organic aspect of living things. Of course he considered the most important of these to be questions related to the psychological and spiritual dimensions of human experience. The individual is autonomous, Chang said, in his attitudes toward morality and values, personal relations, social issues, and ultimate religious belief. By autonomous, Chang meant two somewhat different things: first, that man has "free will"; and second, that no scientific judgment is possible about the truth or falsehood of man's beliefs concerning such subjects.

To explain why this is so, Chang relied heavily upon

European idealist philosophers and the biological "vitalists." He maintained that organic life possesses some essential quality which sets it off irrevocably from "dead matter" and makes it resistant to the essentially quantitative methods of scientific analysis. Bergson, he thought, had demonstrated that this uniqueness of life derives from the dimension of time, which conscious beings perceive as a kind of nonquantifiable "duration." [6]

Chang's argument involved him in a number of controversial, specific assertions about science, many of them derived from the doctrines of the biological vitalist Hans Driesch (who was also known to have influenced Bergson).[7] His theory selected what was deemed amenable to scientific analysis by subject matter (the inorganic and the "material") and led to an attack on the idea of sociology: departments of knowledge which discuss the activities of man — history, law, economics, psychology, even some aspects of biology — are "spiritual" (*ching shen*), inherently inexact, and will never be true sciences.[8] The organic uniqueness of a human being precludes any aspect of his individuality being abstracted for the purpose of scientific classification or quantitative analysis. Darwinian biology is only a theory, and that theory is challenged by many biologists in Europe and is unlikely to attain the status of a true scientific certainty.[9] Finally, Chang held science responsible for an all-embracing mechanistic cosmology, from which human free will is absent, and in which all men's acts and choices are presumed to be governed by inalterable scientific law. Moreover, the natural outgrowth of such a cosmos has been the corrupt industrial civilization of the West.[10]

Much of Chang's thesis may be traced to European sources and should be seen against the background of the contemporary Western controversy over free will and determinism. In the eighteen and nineteenth centuries, advances in physical science had led men to describe the world of nature as operating in a completely predictable fashion, in accordance with fixed and certain laws discoverable by science. After the 1850s advances in biology and psychology appeared to prefigure the imminent discovery of similar laws governing the human animal and society as well. Idealist philosophers like Eucken and Bergson were solidly ranged against the "deterministic world view," which they felt was the consequence of acknowledging these claims of science. Their anxieties arose in large part because they looked at expanding fields of scientific discovery in terms of perennial Western philosophical controversies. Scientific laws were assumed to cause "necessary" consequences in a logical sense which went back to Aristotle: to supply antecedent conditions without which such consequences could not occur and given which they must occur. When transposed to the psychological realm, such an idea of "necessity" undermined the concept of free will, with implications that were shattering to ideas of human responsibility for action and the autonomy of the human psyche.

Westerners like Bergson were perplexed by the problem of science conceived in this light, seeing a basic conflict between science and consciousness as the modern form of the old philosophical issue of free will and determinism. Chang took over these ideas, along with the problems buried in them. But Chang also subjected his acquired doctrines to interesting shifts in interpretation which proposed relations between the doctrines and issues of traditional Chinese philosophy, with important consequences for the course of later discussion in the debate. The European concept of free will emphasized the meaning of action and of responsibility for action. Chang emphasized that free will was essential for human *knowledge* of truth and value. "Why do men possess a moral sense of right and wrong, and a sense of responsibility, and sometimes wish to repent? The cause of a moral decision is to be found in the inner heart and not externally, and such a mental decision derives itself from free will and not determinism." [11]

In this way Chang's discussion quickly transferred attention away from a European problem — free will — whose philosophical foundation (the nature of "necessity") had little meaning for him, and which educated Chinese associated, when they thought about it at all, with the purely fatalistic Taoist idea of "destiny." Instead he turned to a problem which had rich import in a long Chinese philosophical tradition: the source of man's knowledge of the deepest "principles" which govern his life and the universe. Ultimately what Chang defended against science was the neo-Confucian idea of intuitive knowledge and the values associated with it. In particular Chang was attached to the doctrine of "mind" of the Lu-Wang school of Sung metaphysicians, according to which a contemplative, inner intuition (*liang chih*) is held the source of all knowledge of "*li*," [12] and the capacity for right action is deemed a direct ancillary to the perfection of that knowledge.

Chang concluded that only by relying upon the free, intuitive, "inner" side of his nature, which does not depend upon externals, can man learn the principles of self-cultivation which will lead to a spiritual culture.

A second major shift in the interpretation of European doctrine came through Chang's use of the idea of a deterministic cosmos, which originated in European traditions of materialistic metaphysics. For Europeans the "universe of inalterable law" was usually imagined in terms of a strict metaphysical naturalism, but Chang Chün-mai followed the pattern of discussion of Eastern and Western civilizations by associating this cosmic portrait with all the complex social and economic phenomena of industrial European culture. He and other debaters took over European rhetoric about the mechanistic universe, but they interpreted it as an expression of Western civilization in particular. What disturbed them was not the logic of necessity but the technological and social world of industrialization; they saw the routines of machines as the things which were mechanistic and the restraints of urban and factory life as posing the limits of freedom.[13] And so, although Chang had begun his essay stating that science could not "govern" or dictate a view of life, he did not end it concluding that science is therefore morally neutral, or even, as some Europeans did, that reliance on science creates a world "without value." Rather he thought that science generates a mechanistic cosmology and that society must mirror the shape of this cosmic pattern and so be infected with the selfishness, greed, and violence of a demonically conceived social Darwinism. The European idea of a mechanistic universe took on the all-embracing aspect of a traditional Chinese system of nature — encompassing natural, ethical, and social principles in an interlocking and mutually interacting whole.

To Ting Wen-chiang it seemed that Chang had no understanding of the nature of science and that philosophically he was relying upon the obscurantist tradition of European idealism. Moreover, on the Chinese side, Chang's arguments appeared as attempts to save from iconoclastic attack a reactionary neo-Confucian metaphysic and aspects of the old culture which could be defended only by irrational appeals to intuition. To Ting, Sung metaphysics had already been discredited by the Ch'ing empiricists, and if unaccountably allowed to revive in the twentieth century, it threatened to turn Chinese society away from scientific progress altogether. He had little interest in simply confuting Chang, a personal acquaintance with whom he often debated in private conversation. His real audience was the younger generation. "The aim of my essay," he wrote, "is not to save Chang Chün-mai but to awaken the youth before they succumb to the same malady." [14]

Ting argued that all the reality that man can know anything about is part of the natural world and that all true knowledge of this world is scientific knowledge. Both Ting and Chang accepted the idea that there is a stark antithesis between knowledge obtained by scientific reasoning and intuitive knowledge, which between them exhaust all the possible ways of knowing anything. For Ting, scientific knowledge was identified with rationality itself, and he simply declared that intuitive knowledge is an illusion. To entertain the idea that world views might lie outside the range of logical analysis will lead to intellectual anarchy:

If this were really so there would be no need to study, no need to seek learning. Knowledge and experience would all be useless. One would need only to "advocate that which arises from the dictates of one's own conscience" (liang-hsin), because views of life "all arise from the spontaneous impulses of conscience and there is definitely nothing which causes them to be such" . . . Moreover there would be no ground for discussing any question at all. The fact that all debates must use logical principles and that they must all have a method of definition are things which Chang Chün-mai's view of life does not recognize.[15]

As for the fact that the diversity of prevailing beliefs among men might be said already to amount to intellectual anarchy, this is a psychological fact about the present, not an inevitable principle. Differences of belief are the result of imperfect knowledge, of prejudice, of emotion or derangement in human beings; but little by little science is revealing facts about nature, including man, which gradually become assimilated into the world views of all rational individuals. Ting's great examples were the Copernican and the Darwinian revolutions, which have dramatically altered man's view of his own nature. He expressed confidence that future discoveries in psychology and sociology would expand man's knowledge of himself, and that it would be a concomitant to the progress of knowledge in general that gradually the subjective element in man's opinions would operate in ever more limited areas. These views closely echoed the language of Thomas Huxley.

What did Ting mean when he said that science can know everything that is knowable? This was his basic thesis, by means of which he attempted to refute Chang's division of reality into matter and spirit and to defend the social sciences and with them science's claim to be relevant to the study of man. His arguments were drawn almost directly from the classical statement of Anglo-Saxon critical positivism, *The Grammar of Science* by Karl Pearson. Pearson had been inspired by hostility to Continental idealism, which he thought formed the chief obstacle to the extension of scientific method to the treatment of human and social questions. He based his case for the comprehensiveness of scientific knowledge on an explanation of the scientific method in terms of an empiricistic epistemology.[16]

We know scientific facts, Ting explained, in the same epistemological sense that we know anything, as a result of a process which takes place in the mind. How, he asked, can something which is known only through the mind be called pure matter? Specifically, Ting asserted that material substance is known only through the medium of human sense perceptions: "Thus any thought, no matter how complex, does not lie outside the apprehensions of the senses (*chüeh-kuan ti kan-chü*). Directly they form the stimulus to thought; indirectly they form the original subject matter of thought. But the trained mind can fly from the first kind of apprehension to the second, analyze them and make associations concerning them, and from direct perception proceed to indirect concepts." [17]

However, sense perceptions, through which all the basic data of science are understood, occur in the brain, and their relation to extended substances is epistemologically obscure: "The material substance of which we are aware is nothing more than apprehensions of the senses in the mind. From perceptions, concepts are formed, and concepts give rise to inferences. Since what science investigates does not lie outside such concepts and inferences, can there be any difference between spiritual and physical science? Further, how can we say that purely mental phenomena are not governed by the scientific method?" [18] Ting copied directly from Pearson a metaphor of the mind's relation to material reality as resembling that of a telephone operator in his exchange to the invisible callers at the end of the line: "Our nervous system is like a telephone system. The brain is a kind of very powerful switchboard operator; the sensory nerves are the incoming lines and the motor nerves are the answering lines. If the switchboard operator is forever locked up in the main exchange and is not allowed to go out and see callers face to face, what method has he of investigating these customers except through their voices over the line?" [19]

What we know, Ting concluded, are the arrangements of reality which our mind constructs out of sense data, but we will never know what the material reality behind sense data really is. Moreover, we will dismiss the question of what it is as lying outside the range of human faculty and thus intrinsically unknowable. To proponents of this epistemological theory Ting gave the English name of "skeptical idealists": "Because they consider that sense apprehension is our only method of knowing physical bodies and that concepts of physical bodies are phenomena in the mind, they are called idealists. Because they acknowledge that they do not know, and believe that men should not ask, whether or not there exist things outside the world of sense perception, or behind the surface of self-consciousness, or what kinds of things physical bodies fundamentally are, they are called skeptical." [20]

Finally, Ting explained how the epistemology of "skeptical idealism" still allowed a philosopher to be confident that human beings all have similar sense impressions which can form the basis for common scientific knowledge. Although other minds are unknowable in reality just as the physical world is, men may argue from their knowledge of their own minds and their perception of other men's behavior that others possess a mental apparatus like their own. It was, he thought, an argument based upon the uniformity of nature, which is the basis for scientific generalizations of all kinds. "I deduce from the phenomena of my own self-consciousness that other men also are self-conscious, which is not contrary to the scientific method." [21] In this way it is possible for the organizations of perception that make up one man's knowledge to be taken as true for all normal men. Ting concluded with the claim that these doctrines represented a "scientific epistemology," and that in spite of superficial differences, basically they were held in common by scientists as diverse as Huxley, Darwin, William James, Karl Pearson, Dewey, and Mach.

Skeptical idealism presented more problems than Ting realized, and the European thinkers he referred to did not fail to trip over the difficulties involved. Ting did not seem to be worried by, or even to be aware of, the radicalism of his epistemology, which left the very existence of a physical world outside the mind in a state of Cartesian doubt. The phrase "skeptical idealism" echoed "critical idealism," a locution in an essay by Huxley which Ting had probably read, wherein Huxley traced this epistemological position to its origin in Descartes and then clumsily attempted to reconcile it with his own confidence in the existence of extended substances.[22] Intent upon destroying the idealist dualism between matter and spirit, Pearson had been willing to posit the logically purer and more extreme version of the doctrine, which Ting in turn adopted;[23] but in so doing Pearson had precipitated an inevitable counterchallenge in Europe, not so much from metaphysicians as from friends of science like John Dewey. They questioned a theory which seemed to divorce the concepts of science (and by implication, of all knowledge) entirely from any world of real, experienced, extended substances to which they might be said to be related, and which said that such concepts are no more than convenient fictions by means of which we organize our perceptions.

These criticisms are most pertinent, but in fact during the course of the polemic in China nobody used them to counter Ting effectively. Chang's rebuttal, where it did not simply appeal to men's intuition of Intuition, relied on other difficult issues of European epistemology. He did not see that Ting divorced all mental phenomena from material substance, but thought that he refused to accept the possibility of a mental reality which was not linked to any material stimulus. Chang called Ting a "sensationalist" and assumed (incorrectly) that his doctrines were essentially identical with the epistemological theories of Locke. Therefore he attempted to refute Ting by bringing up Kant's classical arguments against the British empiricists. Kant, he said, knew that the contents of sense perception would be quite incoherent were it not for a formal ordering faculty of the human mind which conceptualizes them and makes them assimilable. This organizing capacity is internal and depends upon no sensations for its existence. Chang, however, encouraged his critics to dismiss his point when he persisted in describing this formal Kantian epistemological faculty as a kind of intuitive knowledge, which most of his Chinese opponents associated with unacceptable notions of neo-Confucian mystical insight.

Most of the proscientific participants in the debate rejected skeptical idealism as a concession to metaphysics. This was the position of Hu Shih, Wu Chih-hui, and Ch'en Tu-hsiu. To all of them equally the assertion that something is unknowable due to the structure of consciousness itself seemed to leave the way open for men to invoke a metaphysical Unknown looming behind the veil of appearances. There is no doubt that many Chinese tended to think in this fashion. Yen Fu, for example, had found a similar positivist epistemology in Mill's *Logic* a confirmation rather than a denial of his own faith in inconceivable realities inaccessible to ordinary human experience; and he had passed this interpretation along as part of his standard Chinese translation of Mill.[24] Buddhist and Taoist mysticism placed too much emphasis upon phenomenological characterizations of an apparent world of perceived experience and an opposed spiritual reality for such speculations not to seem apposite to individuals steeped in Chinese religious tradition. The proponents of

science countered with a rigid empiricism which demanded that all forms of reasoning have direct reference
to substances in the physical world. In so doing, however,
they as well as the metaphysicians remained insensitive
to the distinctive character of a European epistemological
theory which insisted upon the *conceptual* limits of knowledge. This kind of limitation does not allow anyone to discuss meaningfully knowledge inaccessible to the presumed
structure of consciousness.[25]

Perhaps in the context of the debate Ting's skeptical
idealism did not have to be understood. It served in China
the polemical purposes it had been designed to serve in
Europe: it attacked Chang Chün-mai's division of reality
into the physical world of science and the spiritual world
of mind and intuition by pointing to the common mental
foundation of our knowledge of everything. It made it possible to say that science is not a subject defined by a certain
type of data to which its methods are relevant. Rather the
material of science is coextensive with all objects of human
knowledge, which are reducible to constructs and concepts in the mind. What characterizes science is a way of
knowing, not a set of things known.

In this way Ting came to the fundamental conclusion
that "the omnipotence of science, and its comprehensiveness, lies not in its subject matter, but in its method." [26]
Moreover, he said, as a method it does not yield a set of
deterministic, inalterable and mechanical laws as Chang
had claimed. It was by a new characterization of scientific
law that scientific philosophers in Europe attempted to
dispel the idealists' bogey of the "mechanistic universe."
Aristotelian concepts of causation and necessity, they said,
are not properly applied to scientific laws, which are better
pictured as economical descriptions of regularly occurring
sequences of events. In the words of Pearson, "that a cer-

tain sequence has occurred and recurred in the past is a matter of experience to which we give expression in the concept causation; that it will continue to recur in the future is a matter of belief to which we give expression in the concept probability." [27]

Causes in science mark stages in sequences of events; they do not enforce or originate them. Sequences of sense impressions in nature are not logically necessary like deductions in geometry; a logically necessary element enters into science only because the mind systematizes perceptual routines according to rules which are the intellectual products of man, not the order of the outside world. Theoretical scientific concepts like "gravity," "mass," and "force" are not mysterious entities which provide the motive power for the deterministic universe, but merely convenient fictions according to which we organize our perceptions of experimental data. The order of nature is an order placed there by the organizing logical faculty of the human mind; it is not some set of inherent principles in things determining the course of change.

In this way Europeans denied that scientific laws were in a philosophical sense deterministic. Ting, of course, upheld their views, but he had two difficulties when he tried to present this chain of reasoning to his Chinese audience. One was that though he accepted the conclusion he did not himself understand the arguments leading to it very well; the other was that Chinese opponents — having linked determinism with an attack upon the possibility of intuitive knowledge of moral truth — were in no position to be affected by ratiocinations designed in Europe primarily to clarify the logical status of scientific laws. The result was a good deal of confused discussion on both sides.

Ting merely reproached Chang for his metaphysician's

habit of making everything into an "immovable system" (Ting used the English word "system"), which led to the grave error of supposing that science is fixed and unshakable and that its rules are unchanging. Newton himself recognized that scientific laws may change and be superseded, Ting claimed, and Ting also seemed to believe that the grip of "necessary laws" could be loosened simply by recognizing that new scientific laws, as he thought, replaced old ones: "What is called a scientific law is a method of speaking clearly about observed facts; if it does not fit newly discovered facts, it can be changed at any time. This is what Mach and Pearson mean when they refuse to acknowledge that scientific laws are necessary (*yu pi-jan hsing*)." [28]

This explanation did not get at the root of the philosophical error in attributing a deterministic kind of necessity to scientific laws; that error derives from confusing the logical nature of a scientific theory about the world with some notion of the "metaphysical nature" of the real world itself. Moreover, Ting was misled in thinking that scientific theories supersede one another in the way empirical generalizations must be altered upon the discovery of new facts. These problems derived from basic difficulties in Ting's understanding of the nature of a scientific theory. However, whether Ting was right or wrong did not matter; his Chinese adversaries dismissed his arguments as irrelevant. To them necessary law was simply identified with something like "truth" of any kind. Chang Chünmai said that attempts to deny the necessary character of scientific law were no more than verbal equivocation; Liang Ch'i-ch'ao was disturbed because if scientific laws were not necessary he could see no legitimate ground for using them as a basis for belief at all.

"The omnipotence of science . . . lies not in its subject

matter, but in its method." Since so much focused upon this assertion, Ting did his best to give his readers a formula for characterizing the scientific method. It relied heavily upon the practices of the natural history sciences which he knew personally, like geology. The fundamental operation of a science is classifying, and its fundamental goal, description: "What we call the scientific method is no more than taking the phenomena (*shih shih*) of the world and analyzing them into kinds and seeking their order. Once they are clearly analyzed into ordered kinds, we think of the simplest and clearest words to form a generalization concerning these phenomena, which is called scientific law." [29] Furthermore, the basic logical method of science is that of induction, amplified of course by appropriate infusions of deduction as well.

Ting obtained his definition of a scientific law directly from Pearson; for an account of scientific logic he referred his readers to the standard English text of the time, *The Principles of Science*, by W. S. Jevons, which analyzed scientific reasoning entirely in terms of inductive and deductive procedures. In addition, he liked to recommend an essay by Thomas Huxley, "The Method of Zadig," which the Englishman had written in order to explain to the British public the methodological foundations of paleontology and Darwinian biology.[30] The Zadig of Voltaire's tale practiced model simple induction, being a Persian Sherlock Holmes who could infer from tracks in the sand, fallen leaves, and marks upon pebbles the size and shape of a horse and dog which had passed that way, even though he had never seen them.

Basically then, Ting held an idea of science which treated scientific laws as universal empirical generalizations which are logically constructed out of reports of experiments in a strict inductive way. Generalizations de-

pend upon confirming instances of fact. The entities
analyzed in experiments and described in generalizations
are things as they are known by ordinary experience —
rocks, animals, clouds, and so forth — and the language in
which they are described is understood exactly as is the
language of ordinary experience. This analysis of science
corresponded quite closely to the kinds of things Ting
did as a geologist, or even to what Darwin did as a
biologist; moreover, Ting was able to use it successfully in
the long sections of his essay in which he attempted to defend the scientific character of work done in "spiritual
sciences" like psychology and sociology. He pointed out
how statistical methods hold the key to classification in the
social sciences and how generalizations based upon them
have the lawlike character he attributed to universal empirical generalization. Moreover, this was an idea of science
which enabled Ting to praise freely as scientific instances
of inductive reasoning from the most diverse sources. He
detected scientific reasoning in works of literary and
economic analysis like Hu Shih's study of *The Dream of
the Red Chamber* and Keynes's *Economic Consequences of the Peace.*[31] It enabled him to see a close ap-
proximation between the *k'ao-chü* techniques of the
Ch'ing dynasty classical scholars and the scientific method.

To the contemporary Western reader what is absent
from Ting's account is the idea of a true scientific theory
of the kind that lies at the heart of basic sciences like
physics and chemistry. Basic physical laws, like laws of
motion, are logically quite unlike universal empirical generalizations, which can be challenged by the production
of counterexamples. Facts that do not fit such basic laws
simply lie outside the scope of a scientific theory; they
do not necessarily refute it. The relation of such theories
to experimental data is, moreover, in the words of one

philosopher of science, most like that of maps or models to the territories or structures they portray — something logically quite distinct from the relationship of experimental facts to empirical generalizations inducted from them. Finally, such laws use theoretical concepts as the basic building blocks of their models — concepts which derive their validity not from any direct relationship they bear to observed experience, but from their function within the model itself.[32] It was not surprising that Ting remained insensitive to the theoretical side of science. Why he did so can be answered only with reference to the state of the philosophy of science in Europe in the early twentieth century and to his own Chinese intellectual heritage.

For reasons of their own, Huxley, Pearson, and Mach, the European thinkers who had provided Ting with his scientific education, all depicted the scientific method as resting upon inductive logic and as being descriptive in nature. The deep-rooted empiricism of the English tradition had led generations of thinkers to assume that the logical model for science must be the Baconian one of induction and deduction, and thus to formulate rules for all science which did apply to the natural history sciences (botany, paleontology, geology) more adequately than to the psysical ones (chemistry, physics). As long as their philosophical adversaries were the idealists, European scientists clung to their characterization of the scientific method as descriptive generalization because it helped them to avoid misleading attributions of causation to scientific law. Pearson and Mach may have misrepresented the logical structure of scientific laws, but they were aware that science uses theoretical concepts which, though controlled ultimately by experimental data, have no direct relationship to them. They did not, however, have a coherent, logical account of the relation of theory to experiment, nor were they anxious to emphasize science's nonempirical side in their debates with idealists who wished to infuse the concepts of science with metaphysical meaning. Ting was encouraged in his empirical view of science by the form of their polemic, and because he happened to be a geologist, little in his own research experience hinted that these theories might not always match practice. He learned to object forcefully when men interpreted science metaphysically, but he did not learn the difficult concept, still not completely understood, of a nonempirical theory with a strict function in an experimental framework.

A strictly empirical characterization of the scientific method was also congenial to Ting as a Chinese appreciator of those Confucian scholars whose work was impregnated with a nominalistic concept of fact and a program to "seek principles in things." Facts were linguistic descriptions of concrete, perceived things, and a narrowly inductive style of reasoning had been the method of finding truth intuitively used by the "solid scholars" of the past. Like these Ch'ing scholars, Ting could see no alternative to this mode of thinking other than untrammeled speculation. The antithesis between science and metaphysics seemed to be an exclusive one between common sense practical reasoning controlled by straightforward evidence of the senses and wild theorizing as unrestrained as the imagination itself. Between them, these represented the only two styles of thinking possible. This had been the usual way of describing the old controversy within neo-Confucianism between the metaphysicians of the Sung and Ming stripe and the classical scholars of the Ch'ing, and the participants in the 1923 debate specifically related their views to one side or the other of that ancient dialogue. Chang relied heavily upon the idea of intuitive knowledge as presented by Lu Hsiang-shan and Wang Yang-ming; Ting polished the old invective of the angry "solid scholars" when he attributed the decadence and fall of the Ming dynasty to the fashion of empty speculation which dominated officials of the old imperial court, and he implied that modern metaphysicians would provoke similar calamities if they obstructed scientific reforms.[33]

Thus Ting's concept of science had been formed by a native intellectual tradition with which he self-consciously identified himself and by European philosophical accounts of science which, for their own reasons, happened to reinforce rather than counteract certain elements in Chinese intellectual patterns. Disciplined by long years of scientific practice and a comparatively sophisticated understanding of contemporary scientific thought in Europe, Ting was able to make the fundamental distinction between science as a logical methodology and science as a metaphysical system of nature, and he had rested his case upon a characterization of science as a way for discovering truth and for learning to manipulate man's environment. But he had been able to give only a narrow, inductionist account of what the scientific method consisted of. His fleeting references to "logical rules, methods, and definitions" were not much help to readers who had never studied individual sciences and found nothing in their intellectual background which prepared them to distinguish between a logically controlled scientific theory and pure speculation. In the debate many enthusiasts for science were no different from the metaphysicians in treating what they understood to be scientific theories as if they were speculative ones. This tendency was especially evident in the essays supporting science by Hu Shih, Wu Chih-hui, and Ch'en Tu-hsiu.

Hu Shih and Wu Chih-hui expected of science principles from which one could explain the nature of the cosmos and the nature of man as an ethical and social being — a metaphysical system of nature to replace the religious world views of tradition. When the metaphysician Chang condemned science as the generator of an abhorrent conception of the universe, Wu and Hu simply welcomed science as the inspiration of a new systemic philosophy.

The first man to volunteer a new materialist bible was the aging iconoclast and Kuomintang militant Wu Chihhui, who had made his reputation as a witty popularizer of evolutionary, anarchist, and anti-Confucian doctrines in journals like *New Century* and *New Youth*. Wu was proud of his atheism and of his position as a defender of "materialist civilization," which alone, he claimed, is capable of producing a worldwide utopia. He considered that the central controversy of the science and metaphysics debate had not been over the laws of science, but over the conservatives' claim that materialist culture leads to war rather than to progress, abundance, and happiness. The truth was that "there will be no war only when material culture has progressed to an unimaginable degree, and there is worldwide, compulsory higher education." [34] Moreover, Wu spoke as if this future world of miracle-working machines would be an earthly manifestation of a cosmological order — an order which he outlined in his widely read essay "A New Belief and Its View of Man's Life and the Universe."

Appropriately, Wu's essay opened with Being: all that is and is not, he said, ultimately is One (*i-ko*), encompassing all things, embracing all contradictions. In addition, the One is composed of "living substance": it is materialistic because it is substance, and it is living because science teaches us that all substance has that salient characteristic of life, sensation: "A stone has matter (*chih*) and force (*li*), and force is manifested in the attraction of chemicals, which is no different from sensation . . . When in fact did the stone in a latrine, the rosebush, the fly and man ever have sensations, a mind, a soul? It is only because the structure and response of their matter and force are different, that is all." [35]

Having established an ontology, Wu moved on to Creation, the process by which the One is transformed into the Many:

In the beginning without beginnings, a laughably chaotic, indescribable monstrosity living in the district of nothingness said to itself, I am bored to death! Who wouldn't be bored to death day after day, year after year, forever, with all this unhearing, unseeing, unfeeling, unsmelling, untouching! . . . Then in the twinkling of an eye, unconsciously, it burst into smithereens. This fragmentation could be said to be in accord with its will — what is called the search for and the finding of benevolence. Instantaneously there arose thousands of universes, or in other words, millions and millions of "I" (*wo*). Its method of transformation was very simple. It was no other than to take so much "unthinking" mass, possessing matter and force, and to make it into whatchamacallits. So many whatchamacallits were made into particles of electricity, and particles of electricity were made into atoms. Atoms became sun, moon, stars, mountains, rivers, grass, trees, birds, beasts, insects and fish. If you are pleased to call it continuous creation, you may; humorously to dub it an illusion of the mind is all right, too. To conclude, it has not finished its transformations even today, and moreover it seems that no kind of thing is considered so satisfactory that it is protected from further transformations forever more. This is my view of the universe.[36]

Next, he announced that the ultimate tendency of the eternally changing universe is to move toward "truth, beauty and goodness" — a goal which, though never actually attainable, is better approached through a "Western" life of striving for material and social improvement than through an "oriental" fatalism and self-abnegation. Man himself is an intelligent, playacting animal, whose role in the cosmic drama is infinitesimal and evanescent, but who has learned to play his part according to the rules of his own physical nature.

Finally, Wu set forth the norms of a morality, extrapolated from what he conceived as man's natural instincts for food, sex, and friendship. These instincts, he thought, provide the biological basis for all conduct, from the highest to the lowest human acts. The place of food in nature requires man to earn his bread by his own labor, to abhor theft, and to consider it honorable to work to provide for others. The straitjacket of marriage distorts the natural role of sex, but in their free sexual relations couples should maintain the standards of chastity which were in Chinese practice the ideal of respectable, Confucian husbands and wives. To fulfill his natural instinct for friendship, man must rationally practice the Mencian cardinal virtues of sympathy, shame, humility, and justice. For Wu, man in the state of nature conducted himself much in the manner of an upright, petty Confucian gentleman.

What did science have to do with this cosmological fantasy? For Wu, nature was not the indifferent, neutral multiplicity of things which presents itself to the scrutiny of scientists; Wu conjured up a prescientific, purposive Nature, potent with destiny, which embraces everything there is in the universe and carries mankind in its bosom into the mists of the evolutionary future. Professions of atheism and invocations of "matter" and "force" did not rid Wu's system of nature of its metaphysical content;

they simply embroidered his teleological naturalism with specific examples of metaphysical interpretations of ideas in science of just the kind which scientists in Europe had been criticizing. A few misunderstood concepts in physics and biology, selected precisely for their vulnerability to mystical interpolations, had been taken as providing the "scientific" cornerstones for a set of speculations thoroughly in the spirit of Buddhist and Taoist cosmology. It is ironic that Wu, the man who boasted that he had read no traditional Chinese books for twenty years, was particularly ignorant of the temper of Western science and, of all the participants in the debate, was the one whose contribution was most reminiscent of the themes of Chinese religion. Wu's cosmology differed from ancient Buddhist-Taoist models chiefly because it rejected the idea of cycles and claimed that the movement of eternal flux is progressive.

Wu sensed that there were limits to the scientific character of his world view, but he did not trace these to the real problems: his idea of nature, his interpretation of scientific concepts, and his purely speculative method, all of which followed from his demand that science provide a world view in the first place. He admitted merely that certain of his statements had not yet been empirically verified, but he thought that by refusing to fill in gaps in existing knowledge scientists suggested falsely that scientific knowledge is in some way restricted in principle. Wu wore his own metaphysical ghost with panache. "My metaphysical ghost has been baptized by the god of science," he said. "Because he strives to be of assistance to science, this metaphysical liar need not get a beating from Ting." [37]

"A New Belief" owed some of its popularity to Wu's racy and pungent style, modeled on the colloquial of China's classical adventure novels. Readers were startled with scatological slang and were treated to well-aimed gibes at Chang, Liang Ch'i-ch'ao, Hu Shih, and other academic luminaries. Under the artful guise of plain speaking, Wu compared himself to "an old man, leaning against the woodpile, warming himself in the sun," and he played the role of village wiseacre puncturing the pretensions of "those formula matching, phrase juggling, addled and swell-headed philosophers!" [38] But further, the substance of Wu's argument appealed to many. This is indicated by Hu Shih's estimate of Wu Chih-hui as representative of the temper of mind of the Chinese Renaissance and by his praise of "A New Belief" for its "clear thinking" on the issues of the science and metaphysics debate.[39]

Because Hu Shih was prevented by illness from participating in the first round of exchanges, he attempted a résumé of the issues in an introduction to an anthology of the principal essays in the polemic. Wu Chih-hui alone, he said there, had made it clear that the reputation of science depended upon Mr. Science's ability to produce a comprehensive, reliable, scientific view of life. In Hu's opinion, it was the duty of the defender of science to venture a scientifically plausible explanation of the universe, its contents, and of man, and not to retreat, as Ting had done, behind the "idealistic" formula that some things are unknowable. Therefore a proper debate ought to center around attacks upon and defenses of a postulated scientific world view. For his own part Hu offered for public consideration a portrait of the scientific cosmos which he thought to be a slight improvement on Wu's version, one which was "based upon the great hypotheses of three hundred years of scientific knowledge":

1. Astronomy and physics teach man that space is infinitely large

2. Geology and paleontology teach man that time is infinitely long.

3. All science teaches man that change is natural to the universe and to all things, and that it is self-generated, needing no supernatural God or Creator.

4. Biology teaches man of the waste and cruelty of life's struggle for survival, and hence enables man to understand that the hypothesis of a "benevolent" God cannot stand.

5. Biology, physiology and psychology teach that man is only one kind of living organism, differing from other kinds of living organisms in degree, not in kind.

6. The biological sciences and anthropology, ethnology, and sociology teach man that living organisms and human societies have had an evolutionary history and are subject to evolutionary causes.

7. Biology and psychology teach man that all psychological phenomena have causes.

8. Biology and sociology teach man that doctrines concerning morals and manners are subject to change, and that the causes of these changes may be investigated by the scientific method.

9. Modern physics and chemistry teach man that matter is not dead but alive, not static but dynamic.

10. Biology and sociology teach man that the individual — the "small self" — is mortal, but that mankind — the "great self" — is immortal and not subject to decay. They teach man that "life for all mankind and for all time" is religion, the highest religion, while the religions of a "heaven" or a "pure land" planned for the individual after death are religions of selfishness.[40]

Hu referred to his ten propositions as the "Decalogue" of the "naturalistic universe," and he concluded with a rhetorical affirmation of the human right to values in such a cosmos:

In this naturalistic universe, limitless in space and time, this two-handed creature who averages five feet six inches and no more than a hundred years — man — is truly a petty and insignificant creature. In this naturalistic universe, where the operations of nature are ordered and things change according to inherent laws, where laws of cause and effect govern all his — man's — life, and the ruthless struggle for survival whips him on in all his deeds, the liberty of this two-handed creature is extremely limited. But within the naturalistic universe this small two-handed creature still has his proper place and his proper value. Using his hands and large brain, he even can fashion many tools, devise many techniques, and create a culture. Not only does he tame many beasts, but further, he can investigate the natural laws of the universe and make good use of these laws to yoke the operations of nature; today he even can call upon electricity to propel his vehicles and send his letters. The advancement of his intelligence leads to the increase of his power; but the advancement of his intelligence also broadens his sympathies and increases the powers of his imagination. He has worshipped inanimate objects and animals and has feared ghosts and spirits, but today gradually he is leaving behind the age of his immaturity. Now gradually he is coming to know that the vastness of space only increases his feeling for the beauty of the universe; the duration of time only heightens his awareness of the difficulties his ancestors had in creating a heritage; the order of the operations of nature only increases his capacity to control them. Even the laws of cause and effect embracing everything need not shackle his freedom, because on the one hand the action of causal laws enables him to seek causes from their effects and from effects to infer causes, to explain the past and to predict the future; and on the other hand it enables him to employ his intelligence to create new causes and to seek new effects. Even the idea of the struggle for existence need not make him into a cruel and unfeeling beast, but may yet enlarge his sympathies for his fellow creatures . . . In sum, this naturalistic view of man's life is not for a moment without beauty, poetry, or moral responsibility, nor does it lack in opportunities for the full use of "creative intelligence." [41]

There is practically no difference between the portrait of the scientific universe Hu offered and the one Chang Chün-mai was objecting to. It is the nineteenth-century Great Machine, governed by inexorable laws of cause and effect which determine the course of nature, animate and inanimate alike. Hu tried to defend free will by suggesting that man can interfere in the operations of causation and somehow presumably choose his causes and, with them, their effects. He did not try what Ting tried: to declare that the entire concept of causation implied by such a formulation was what was at fault. Wu and Hu Shih understood the nature of science much as Chang Chün-mai did; they differed only in their satisfaction with the result.

For all of them scientific hypotheses were presumed to establish a cosmological system, embracing all nature and man, and scientific propositions were used as springboards for speculating about what that system might be. In each case the result was a set of a priori formulations designed to answer traditional "great questions" about human nature and destiny. They demanded that science provide them with a philosophical theory, and they thought that a philosophical theory had to be a quasi-religious view of the universe. Hu had adopted the furniture for his view from Western sources, from a positivism in the spirit of Comte; Wu's imagination created a phenomenological universe of continual change and transformation as well as a living Nature impregnated with purposes and responsive to man's intentions of a kind that had deep roots in traditional Chinese cosmology.

However, Hu Shih and Wu, far more than Ting, became the popularly accepted spokesmen for Mr. Science. When in the end they fell under attack, their most formidable adversaries proved to be yet another group of men who interpreted science as the generator of a systemic philosophy — the Chinese Marxists. The successor to the science and metaphysics debate was a sprawling controversy on the subject of the Marxist dialectic, which filled the journals in 1929 and 1930. This time a narrow victory of public opinion was awarded the side of "historical materialism." In 1923 only Ch'en Tu-hsiu offered a brief Marxist commentary on the Ting-Chang controversy, but it indicated the type of thinking upon the issues of science and metaphysics which would become increasingly popular.

For Ch'en, who was one of the founders of China's Communist party, it was obvious that a man's view of life, like all psychic phenomena, must be a product of the materialist forces of history. "Only objective material causes can change society . . . explain history, or account for a view of life." [42] He criticized Hu Shih for including among the possible agents of causation epiphenomena like human ideals and intentions, religious beliefs or social customs, and thus for refusing to understand that such things are themselves ultimately determined by the forces of the physical environment. As for Ting, his talk of "skeptical idealism" simply betrayed him into the hands of the metaphysicians, who would be delighted to endorse the thesis that there are things in the universe science cannot know.

Ch'en Tu-hsiu's notions of *matter* had no more to do with the hypotheses of science than did those of Wu Chih-hui. For Ch'en matter was an ultimate force which embodied the course of human history, generating out of itself all culture, society, every human idea and feeling. Like most Marxists, Ch'en was convinced that dialectical materialism was scientific, since the rules which govern physical substances can be understood by empirical observation. But in fact, as is well known, the Marxist system preconditions the spirit in which one interprets social and physical observations; in no way does empirical observation alone supply an adequate verification for the dialectic, which is known to be true by a special nonempirical historical intuition. Like Hu Shih and Wu Chih-hui, the Chinese Marxists also claimed that the postulates of science lead to a systemic world view which they had actually reached on a priori grounds.

Throughout the debate both scientists and metaphysicians returned constantly to the question of how science may be related to moral value. The original question presented by Chang Chün-mai, "Can science govern a view of life?" had been interpreted by all sides as meaning, "Can science lead us to knowledge of moral truth?" The creators of speculative scientific systems were in the happy position of being able to import into their naturalistic universes the human spiritual ideals which happened to be most precious to them. Indeed this contributed to the popularity of their views, for on all sides it was accepted as unthinkable that a set of ideas, especially one so fundamental that it was being considered as a guide in the affairs of state and a basis for the education of youth, could fail to have implications for moral choice.

Like the others, Ting also could not help feeling that it was impossible for science to play the role of intellectual guide and yet remain morally neutral. Without departing from his focus on methodology, Ting suggested two ways of reconciling science and value. First he invoked an ethic of rationality and truth-telling — virtues which could be imagined as a consequence of thinking in a scientific way, and which most people also associate with the moral ideals of trustworthiness and responsibility. "The daily search for truth and the elimination of dogmatism will not only give the scientifically educated man the capacity to seek right principles (*ch'en li*), but also a sincere love for them," he said.⁴³ He also claimed that the basic questions which divide mankind are not those of ultimate ends, but those of means; thus the scientific method, diffused through education, should teach mankind how effectively to achieve the universal goals of peace and social harmony.

But here Ting was not only saying that sound scientific choices were right and good choices, but also that the great goals he spoke of were themselves facts of nature. Demanding an allegiance to facts as scientific reason discloses them, he moved irresistibly in the direction of seeing the value that happened to be most precious to himself — human altruism and capacity for social self-sacrifice — as an attribute of the species. So he said all living creatures shared a biologically rooted "religious instinct," arising in nature out of the evolutionary lottery and perpetuated because of its utility for the survival of the species. When it came to the crucial issue of ethics, Ting was always a social Darwinist. By trying to make ethical attitudes seem the inevitable consequence of scientific modes of thought, he resembled his adversaries in unconsciously allowing normative propositions to assume the status of scientific fact.⁴⁴

A belief in the close relationship between science and value also lay behind the arguments over the "spiritual East" versus the "materialist West" which many readers and participants imagined to be the central issue in dispute. Provided with accounts of two opposing systems of nature, the scientific and the metaphysical, they imagined that each was like some cosmic mold which, imprinted upon the world of men, left there its characteristic stamp. Metaphysicians were horrified by a vision of a society unmoored from traditional standards of right and wrong and enmeshed in unending struggle and conflict for the sake of gain — presumably the human lot in a mechanistic universe. Their opponents, implicitly assuming (with an ironical bow to Confucius) that evil is the result of deprivation, in anticipation embraced the industrial genie which was to wipe out the sufferings of the past and produce a brave new world of comfort and abundance, and consequently, of virtue. In the eyes of the metaphysicians World War I was the living verification of their fears, yet to the scientists it was little more than a broken cobblestone in the long path of progress.

Again Ting attempted to distinguish between science as a methodology and science as a kind of system of nature — in this case associated with the social systems of the nations where scientific knowledge had been widely applied in the form of industrial technique. He did so by indicating the gulf that lies between science and technology. It is a mistake, he pointed out, to think that science, which is the search for truth in the laboratory, is responsible for the uses put to its discoveries in factories and on battlefields. Rationality guides the scientist in his pursuits, but desire for wealth motivates the industrialist. As for diplomats and politicians who precipitated World War I, most Chinese did not realize that the influential upper classes in Europe were products of a classical and religious education almost as far removed from the scientific ideal as was a Confucian education in China. Ting felt that Europe's troubles actually arose from her failure to extend the spirit of science to her thinking on social and political questions.[45]

The other advocates of science were pleased by Ting's analysis of the causes of World War I and by his characterization of European society, but neither they nor Ting himself carried his distinction between science and technology very far. Whether or not they realized it, that distinction suggested a gulf between science (which produces theories about the natural world) and social choices (where these theories are used for human ends) that none of them was prepared to face. The scientists from a positive point of view, like the metaphysicians from a negative one, insisted that science was relevant to the highest moral purposes of man. Ting, in seeking to free China from "metaphysical ghosts," did not presume to supplant them with a science that was ethically neutral.

Whatever its deficiencies, the science and metaphysics debate did represent the first serious attempt in China to discuss issues of Western philosophy. Such an attempt was possible because of the increasing knowledgeability of a generation of Chinese with more than a superficial exposure to European thought. It seemed worthwhile because problems of science and of social changes resulting from the industrial revolution had in fact become worldwide by the 1920s, and the Western philosophical issues which touched upon them had relevance as immediate in Peking as in London. The polemicists did, however, remain tied to many traditional Chinese intellectual patterns, which influenced their selection and interpretation of European ideas and made the discussion more an example of cultural confrontation than a contribution to the problems of Western philosophy by individuals who happened to be Chinese.

Naturally it was difficult to present complicated issues of European philosophy to the Chinese public, particularly when men also wished to relate these issues to other ideas — in ways equally complicated and often quite different — which had played a perennial role in native intellectual tradition. Possibilities for confusion abounded. Few of the participants had seriously studied particular sciences; most considered themselves simply as nonspecialists attempting to integrate ideas they had received in the course of a mixed Chinese and occidental education. It was easy to misinterpret European doctrines often picked up at second hand and without sufficient understanding of the meaning supplied them by their European historical and intellectual context. The attempt, within a framework of European concepts, to satisfy standards of philosophical meaning and relevance provided by Chinese tradition was a potent source of trouble. Efforts to claim identity between individual European ideas and somewhat related Chinese ones were only among the more striking indications of this.[46]

Equally important, moreover, were the difficulties which beleaguered Europeans themselves. Ting and Chang assimilated standing problems in the versions of European positivism and idealism which they adopted. Like Chang, many European conservatives responded to the stresses of scientific advance by attempts to defend traditional thinking with metaphysical barricades. Like Ch'en, European Marxists used misunderstood concepts of scientific logic to prop up their utopian political theories. The "materialistic West" itself started as a European slogan, born out of the World War and repeated in China even by Bertrand Russell. Europe supplied the form for Chinese doubts even as she created the conditions that provoked them, and Europe passed along to the East, in addition to scientific progress, the intellectual difficulties it had already engendered at home. The Chinese revealed the native form which the problem took partly by their selection from among the various themes available. In this regard the most interesting Chinese preference was the one for system building, Continental forms of philosophy over the analytical positivism of the Anglo-Saxon countries.

Hu and Wu's systems of nature seemed more familiar

and comfortable to most Chinese than did Ting's discussion of science as a methodology.[47] When faced with fundamental questions about nature, the average Chinese intellectual of the early twentieth century, whether sympathetic toward science or not, thought that some form of speculative cosmology supplied the *kind* of answer required. Moreover, he believed that a philosophical question at some point had to involve a question about ethics, and he remained insensitive to the internal guidelines to thought provided by Western logical forms, unless these forms were interpretable in a strictly empirical fashion. This being the case, he tended to make the theories of science into systems of belief, using concepts suggested by those theories — he came to social Darwinism before he came to the science of biology, and he talked of the mechanistic universe before he examined the laws of mechanics.

Long before the debate was over, Chinese polemicists complained that the discussion was proving long-winded and diffuse, and the wiser among them concluded that the controversy should be regarded only as a stage in a long educational process.[48] Still, in the end there was general agreement that science had won a victory, of propaganda if not of understanding. Neo-Confucian philosophy suffered a serious blow to its academic prestige, and elements in Chang's version of European idealism lived on only to achieve a final disreputability as part of the ideological underpinning of the New Life movement. Later one observer waggishly tallied up the score as follows: "The sacred names of Jesus and Buddha were used in vain. Much injustice was done to Confucius, Laotzu, Plato and Aristotle; severe wounds were inflicted upon Kant, Hegel and Bergson; mortal blows were dealt to the philosophers of the Sung and Ming dynasties. Karl Marx, Bertrand Russell and John Dewey were singularly lucky in escaping the rain of bombs and shells." [49] Ting, who of the leading participants had displayed the most sophisticated understanding of science, emerged less as a popular hero of science than did Hu Shih, Wu Chih-hui, and Ch'en Tu-hsiu. As we have seen, his "skeptical idealism" drew fire from both sides, but for the wrong reasons. His most important insight, that science should be regarded primarily as a rational methodology, was for the most part simply bypassed.

However, the enthusiasm for science generated by the debate was in itself a creative force in Chinese intellectual life. In spite of their polemical differences, Ting and the other advocates of science were all motivated by their belief that only trained intelligence — and they thought that science alone genuinely fostered this — could devise proper solutions to China's problems. In spite of the widest philosophical divergences, what they shared with all the European thinkers whom they admired, whether Huxley, Pearson, Dewey or Marx, was the demand for a scientific social revolution in which rational techniques would be applied to the problems of human welfare, making possible more humane and efficacious solutions than those hitherto offered through the insights of customary wisdom. This gave their advocacy of science a political and social meaning which made the debate of more than academic importance. They may have used "Mr. Science" as a talisman, but he was a magic charm designed to cast out superstition, conservatism, and blind allegiance to the past, so that man's intellect might be freed to consider the urgent problems which faced him. In 1924 Ting told his friend Teilhard de Chardin that contemporary intellectual conditions in China most nearly resembled those of the eighteenth-century Enlightenment — a rationalistic rebellion against a system of customary beliefs which had been found wanting.[50] It is characteristic of twentieth-century thought the world over that the philosophe's ingenuous, inborn faculty of "reason" today has been recast to conform to the ideal of the scientific method.

Chapter VI
Endeavor: A Political Education

Among all factions of the intelligentsia the May Fourth movement created a new impetus toward political action. The groups that had always emphasized a revolutionary policy — the Kuomintang and, after 1921, the young Communist party — naturally found it a source of inspiration and recruits. But the more conservative nonparty intellectuals like Ting could not help being affected also. These people, whose best known leaders were perhaps Ts'ai Yüan-p'ei and Hu Shih, often wear the uninformative label of "liberals," but their beliefs were as varied as their personalities. What they did share, however, had its importance. Most were foreign graduates, men of established academic or bureaucratic position, who had no strong partisan political identification. In the beginning most of their energies had been devoted to the new culture movement, which was expected to lay the foundation for a reformed Chinese society. "No politics for 20 years!" was Hu Shih's original motto, and when political crises exposed this slogan as a pious hope, the nonparty intellectuals fell back on a limited "problem oriented" approach to reform. Their preference for practical and nonideological forms of political action was given an intellectual patina by being associated with John Dewey's popular doctrine of pragmatism and was reinforced by their personal observation that Western ideologies were proving themselves difficult to apply in Chinese circumstances.

Therefore they attempted to focus public attention upon concrete evils and to keep their respective theories in the background. Finally, they believed that the intelligentsia had a special mission to lead society. In sum, they shared a pride in their independence, in their pragmatism, and, less explicitly, in being in a long tradition of the "educated classes" who were the natural governors of society.

For such men May Fourth was a disturbing challenge: proof that an aroused public opinion had some hope of being a force against official cliques and military power, and also reproof to those natural leaders of society who allowed inexperienced students to take the initiative in a crisis. The new claims of political action seemed all the more compelling when, as they pursued their cultural revolution, the nation was drifting from disunity to anarchy and warlord rule.

Yuan Shih-k'ai's campaign to become emperor had failed in 1916, but its effect had been to destroy the tenuous national unity that depended upon the allegiance of the powerful military viceroys of key provinces. The succeeding administration, a loose junta led by General Tuan Ch'i-jui, faced open hostility from an alliance of southern militarists, who claimed to represent the Republican legitimacy which Tuan had finally shattered when he drove the original "Old Parliament" out of Peking. Tuan also had to deal with the covert rivalries of other northern generals within the junta and with the de facto autonomy of a number of provinces like Szechuan, Shansi, and those of Manchuria. Tuan hoped to build a new army capable of waging a campaign for national reunification and he was willing to finance his ambitions by means of compromising alliances with the Japanese. But his policies only increased the suspicions of his military rivals and contributed directly to the anti-Japanese passions that provoked the May Fourth demonstrations. In the fall of 1920 dissident northern generals of the "Chihli clique," grouped around Wu P'ei-fu and Ts'ao K'un, were strong enough to mount a campaign which overthrew Tuan's government. However, they in turn immediately faced their own challengers. The warlord of Manchuria, Chang Tso-lin, and the armies of his "Fengtien clique" had acted as allies to the Chihli forces, but Chang's own ambitions to expand his power into north China soon made it inevitable that the Chihli and Fengtien factions would clash.

As the civil war deepened, the effect of corrupt politics upon the social order became more evident, and the social evolutionists of the new culture movement heard their doctrines parroted by apologists for the status quo. Although Hu Shih resisted the revolutionary drift of many of his former associates, he could not entirely maintain his former aloofness toward politics. He had to notice, for example, that by 1920 education was already being financially starved by the militarists, provoking professors to demoralizing agitations for wages and students to hectic political demonstrations. Nor could he ignore that the cultural renaissance itself was in danger of being stifled by press censorship laws like those introduced by Tuan Ch'i-jui. The *Weekly Critic,* for which Hu wrote, was an early victim of censorship.

Therefore it was not surprising that in August 1920 Hu and some of his close associates issued a political manifesto which was a plea for those essential civil liberties needed if the ideas of the new culture movement were to continue to circulate. "Originally we did not wish to discuss practical politics," the manifesto said, but

practical politics has interfered with us at every hour and minute . . . Now, when politics oppresses us so intolerably, we must wake up fully and realize that if the people do not take the initiative in politics, no true republic will ever come into being. But if we want to have the people take the initiative in politics, first we must cultivate among citizens an atmosphere of genuine spirit of free thought and free criticism . . . We know that certain basic minimum liberties are the pulse of life to a people and a society. Therefore we earnestly propose them and ask the whole nation to rise and struggle for them.[1]

Beside Hu, signers included Chiang Meng-lin, T'ao Meng-ho, Kao I-han, Wang Cheng, Li Ta-chao, and Chang Wei-tz'u (Chang Tsu-hsün). They asked for an end to police oppression and press censorship, and they demanded guarantees of freedom of association, assembly, and publication, and of honesty in elections.

This manifesto appeared within weeks of Tuan Ch'i-jui's final defeat by the combined forces of the Chihli and Fengtien armies. In the months that followed, Peking itself was at the center of a north China power vacuum, around which swirled the conflicting intrigues of Chihli and Fengtien warlords. The spectacle of military rule, and particularly the fear that north China would be overrun by the universally detested Chang Tso-lin, acted as a further goad to the nonparty intellectuals. In July 1922, almost two years after Hu Shih's civil liberties manifesto, the same individuals, augmented by a number of others, launched a political review, *Endeavor* (*Nu-li chou-pao*), and published in its second issue a new and more sweeping political manifesto. This time there were sixteen signatories, most of them professors at Peking University: Ts'ai Yüan-p'ei, Wang Ch'ung-hui, Lo Wen-kan, T'ang Erh-ho, T'so Chih-hsing, Wang Po-ch'iu, Liang Sou-ming, Li Ta-chao, T'ao Meng-ho, Chu Ching-nung, Chang Wei-tz'u, Kao I-han, Hsü Pao-huang, Wang Cheng, Ting Wen-chiang, and Hu Shih. Seven other ranking members of the

Peking academic world immediately added their names.[2]

According to Hu Shih, the man who first suggested starting a journal and who was one of its most militant supporters was Ting Wen-chiang. It was he who pressed upon his academic friends the urgency of the Fengtien-Chihli war danger, proposed a study group and then a magazine, and devised the method of financing. He was one of its most regular contributors, and when Hu Shih's illness made publication falter, Ting undertook many editorial duties personally.[3] In his own pragmatism, independence, and sense of social role, he typified the dominant temper of the group, and many of the official views of *Endeavor* editorials echoed those which appeared in other places under Ting's own signature.

Paradoxically, Ting's departure from an official position to enter private business in 1921 had also coincided with the beginning of his career in active politics. In a particularly personal essay written in dialogue form for *Endeavor,* Ting reviewed the train of thought which pushed him toward this commitment and also noted some of the psychological obstacles to such a course which hampered the typical nonparty intellectual. Significantly, he placed the counsels of withdrawal, a life of privacy, and the historical view in his own mouth as "a Chinese graduate from abroad." The exhortations to resolute deeds are the part of an unnamed "foreign friend," exemplar of the Faustian spirit of dynamism and struggle attributed to the West.

The dialogue begins as the foreign friend expresses surprise that Ting has gone into private business:

"I replied with a sigh: It is impossible to be an official any more. I have been in Peking for ten years and I always advocated that good men make every effort to take official posts: if many officials are good men, the government will function well. But in the first place, too few people shared my ambition, and many good men were unwilling to take office; in the second place, to be an official was too difficult and good men were unable to master the skills; and, in the third place, after holding official posts, many good men became spoiled. I saw the number of good men in office dwindle day by day, and the government day by day deteriorate. Moreover, I wanted to do my official work well, and was unwilling to accept bribes or to take outside assignments, and so after ten years of work I remained as before — without any savings. Nowadays salaries often are not paid and men are pinched for necessities. Finally, then, I realised: as long as government is not enlightened, there will be no posts which good men can fill and the idea of office will seem rather distasteful. I also saw that we officials who wished to do our work well could spend many years of hard labor to complete a project and then with one order an ignorant and uninformed politician or official could completely destroy it. I realised that the work I did was built upon sand and lacked all solid foundation, and so I changed paths and went into business.

"*A:* Your judgment is certainly correct, and we don't blame you for your action. But I must ask, how can government become enlightened by having officials go into business?

"*I:* It is very easy to talk of making government enlightened, but it is not something that can be done in a day. The present-day misfortunes of our society are that there are too many failures in business and too many deficiencies in education. There is absolutely no foundation for an orderly government. To build up that foundation, there must be economic development on the one hand, and promotion of education on the other. When business and education succeed somewhat, then we may hope that government can be made truly enlightened. In business, of course, I have no direct relation with the government, but indirectly I can provide a livelihood for a number of men — and for myself — and reduce the number of parasites on society. Perhaps this also is a way of fulfilling my duty to my country.

"*A:* . . . If industry and education do not succeed, politics cannot become enlightened: if politics does not become enlightened, can industry and education succeed? . . .

"I was silent for some time, and then replied with an effort: I, too, know well that government cannot be abandoned. So recently outside of business I have been using my spare energy to discuss politics.

"*A:* Ah! Politics is for action. It is not helped by conversation. It requires all one's strength: it is not a pastime for one's spare energy . . . In this age of acute danger and urgency, survival is at stake and there is not a split second to lose. Modern citizens should not be the kind who do not seek to save the nation, but rather want first of all to protect their own rice bowls. Furthermore, what about the argument that even your rice bowls cannot be protected?

"*I:* You say your country has reached a position of acute danger and urgency . . . This is too pessimistic. Our great misfortune is disunity; because of disunity we cannot disarm; because we cannot disarm, our finances are in chaos; but, fundamentally speaking, even our finances are not hopeless. Naturally the present national fragmentation cannot be ignored, but this kind of fragmentation is a common fact in Chinese history: it is temporary, not permanent. Earlier, we most feared foreign enemies . . . Today, international politics have reached a condition of mutual checks; for a while no one can come and actively interfere in our internal government. Since there is no foreign threat, we have time for our quarrels, to battle with each other, and naturally to bring unity out of division, order out of chaos . . .

"*A:* Your words are those of an historian looking back on decades and centuries. They should not be spoken by a citizen of the republic in 1923: . . . such an historical point of view is a strong point of you Chinese, but it is also a weak point. When a country is overrun by bandits, controlled by warlords, administered by bribes — when government orders are not obeyed, domestic and foreign bonds are not backed up, and high and low have no means to live, how can educated citizens abandon their duty and say that simply because there is no foreign peril, it is not a time of extreme danger and they need not worry? . . . What I least understand is the way you are so long-suffering. Really, it is inexplicable to me how educated men can allow themselves to be governed and controlled by a group of illiterates without making the least movement of serious resistance!

"*I:* We can't entirely accept those words. Haven't you seen the student movement?

"*A:* Now that you bring up the student movement, I have something more to ask you! How can it be that the political movement is given over to youths not yet out of school? . . . It is as if the masters of a household attacked by thieves or threatened with fire were to hide, and give the vital work of fighting fire and theft over to adolescents! . . . I am not bragging about the English, but if there is a great movement in our country, the educated people, old and young, all join in. You need only look at the outbreak of the European war, at how Englishmen rushed into the army, trying to get in first, as if they were afraid to be late. That was patriotism — that was courage! Whether or not your national peril today is greater or less than England's at the outbreak of the war, can men fail to feel it at all? . . .

"*I:* I know that we Chinese find it hard to feel it, and to unite together to make a political movement. But the apathy of recent years is actually a reaction. In the last years of the Ch'ing dynasty all people with aspirations were resolute in their hatred of Manchu corruption . . . all men had a common goal in mind: first, to "drive out the Manchus," and second, to "establish popular government." They considered that once the Manchus were overthrown, the government would change into the form of a republic and peace would reign. Since they had a kind of absolute faith in these things, people were ready to sacrifice, ready to shed blood. The experience of the past eleven years has utterly destroyed this faith. The Manchus were overthrown and replaced by others equally bad: before political forms can be thoroughly altered, representative government is already bankrupt. Without faith, the people wander in all directions, not knowing how China can now make good. What is it that calls for sacrifice? What is there to do battle for? No one can answer! First there was disappointment, then there was negativism, and the result of negativism is apathy . . .

"*A* replied: Again you are speaking from an historical point of view. I cannot contradict you. I only have one word of advice: in today's world a pessimistic, apathetic people who live without faith cannot survive . . . A citizen's knowledge and his responsibility, his duty and privilege, are in direct ratio to each other. In general, I feel that returned graduates from abroad are the most accomplished of China's educated class, and also the most privileged members of society. So on the one hand I wish you every success in business, and on the other, I hope you will not forget politics!" [4]

One by one this article challenged the commonest arguments for noninvolvement in politics heard from nonparty intellectuals: their claim that social and cultural reformation must lay the foundation, their appeal to the long forces of history which always seemed to have resolved China's problems in the past, and their repugnance for a frequently frustrating, and even dangerous, career. It might almost be read as a direct rebuttal to Hu Shih. In essence the dilemma was an old one. It had been faced by generations of literati who wondered whether the duty of the righteous man to serve in the imperial government could ever be reconciled with that government's all too frequent ambition to capitalize upon Confucian prestige without submitting to Confucian restraints. Hu, by limiting the intelligentsia to the purely educative role which he believed to be the true starting point for national reformation, had originally hoped to preserve from political compromise the moral authority, and hence persuasiveness, of the intellectuals' platform. Ting denied that aloofness from politics was either an honorable or a practicable position in a time of political corruption, and in 1922 the arguments for action carried the day.

In this mood of crisis a small band of associates set out to try to turn their intellectual prestige into political power. What did lie within their power was *Endeavor*. Printing costs were low in Peking, and a group of fewer than a dozen salaried professors, each contributing 5 percent of his salary for a few months, could amass the resources to put their views before the reading public.[5]

Because they were committed to practical action without endorsing rebellion or insurrection, their political choice was limited: to work with the material at hand. This meant offering cooperation to a reigning military authority and hoping that the presence of a number of prominent intellectuals in the administration would restrain abuses and encourage reforms. Under conditions of power resembling the traditional pattern of imperial despotism, they in fact hoped to hold that relationship to the center of power which was the ideal of the Confucian bureaucracy.

In the summer of 1922 conditions in Peking offered a plausible rallying point. The Chihli militarist Wu P'ei-fu was at that time at the height of his popularity. First, he was the leader of the armies defending north China from Chang Tso-lin; further, in April 1922, after his successful victory over Chang's army in the first Chihli-Fengtien war, he made the restoration of the "Old Constitution" the official aim of his policy. Wu in fact hoped to acquire prestige as the defender of the constitution, challenging the Republican claims of Sun Yat-sen and the unreconciled southerners. In August 1922, soon after the first issue of *Endeavor* appeared, a quorum of the "Old Parliament" returned to Peking, and Li Yüan-hung, the last man formally elected president of the republic by the original assembly, returned officially to his post. In spite of the continued aloofness of the south and of the Kuomintang wing of the "Old Parliament," the delegates and officials gathering in Peking did represent a tenuous link with Republican legitimacy. Finding the much-battered constitutional line the best starting point for reform, the *Endeavor* group accorded its support to Wu.

Its manifesto, published on May 14, 1922, was first of all a summons to the nation's "best elements" (*yu-hsiu fentzu*) to engage in practical politics. It suggested that their rallying point be the lowest common denominator of reforms which all sensible men could agree upon as constituting "good government." Further, it made a number of practical suggestions which represented the consensus of the sixteen signatories as to what the most practical steps toward government reform might be: "We believe that whatever the ideal forms of political organization may be . . . the best elements of the nation must now calmly and matter-of-factly lower their standards to recognize that the objective of 'good government' should be the commonly held minimum requirement for the renovation of Chinese politics today. With this common objective, we must cooperate to make war on the evil forces in the country." [6] The manifesto also offered a definition of "good government": "Negatively, what we call 'good government' is a proper organization which can control and eliminate all selfish, corrupt and non-law-abiding officials. Positively, it makes the fullest use of the machinery of politics for the welfare of society as a whole, it permits complete personal liberty, and it protects the development of individual personalities." [7]

Moreover, it suggested an interpretation of recent history (identical with that in Ting's article just quoted) which could offer a basis for the hope that the intervention of the "best elements" at this time might yield constructive results. The disorders of the past were not the result of force of circumstances, but of some personal failure on the part of those who were the logical candidates for leadership:

Although the deterioration of China down to its present deplorable state has many causes, an important cause has been that "good men take pride in their integrity." "When good men fold their arms, bad men grab and run." For this reason, we deeply believe that the first step for the renovation of today's government is for good men to have a militant spirit . . . Wasn't the new spirit at the founding of the republic due to the fact that the nation's best elements participated in the political movement? At that time many old-fashioned officials ran away to Tsingtao, Tientsin, and Shanghai, where they engaged in commerce and no longer thought of returning to office . . . Later on, good men gradually grew weary of goverment, and, one by one, they left or went into retirement . . . Then the old-fashioned officials, one by one, returned as councilors, advisors, ministers, and vice-ministers. After 1916 and 1917, good men stood aside with folded arms, watching China split up, watching the Anfu club come to power and run amok, watching Mongolia fall, watching the sellout of Shantung, watching the perversities of warlords, watching the nation's growing bankruptcy and humiliation. Enough! Those who are most to blame — the good men — can now rise up! [8]

The *Endeavor* manifesto's practical suggestions centered around demands for a "constitutional," "open," and "planned" government. The first step toward achieving such a regime, of course, must be the healing of the split between north and south. *Endeavor* called for a new peace conference which would negotiate a recall of the national assembly dismissed in 1917. The conference also would negotiate a method for disarmament. Meanwhile, the existing administration in Peking was urged to trim the swollen bureaucracy, to restore appointments by examination, to create a new electoral law providing for direct suffrage, and to provide a public and balanced budget.

In their manifesto and in the discussion of it which filled the journal's pages during the next few issues, the group repeatedly emphasized the theme of unity. Rejecting ideological discourse, they reiterated that China's real enemies were war, anarchy, and corruption. All men of good will, whatever their party affiliation, were asked to unite behind a program of disarmament, unity, and financial reorganization. "I personally believe that the only hope today is for all to adopt an historical point of view, and to recognize the futility of earlier quarrels," Hu Shih wrote. "Today's greatest need is for the propagation of these very commonplace public objectives." [9]

In this way the group attempted to make its objectives broad enough to attract a wide spectrum of support, yet definite enough to be a plausible basis for a program of militant action. In an article defending the manifesto against its critics, Ting explained the group's relationship to political action and further described the principles which united it. *Endeavor* readers had offered two criticisms of the manifesto. If you really want action, they asked, why don't you form a political party, or alternatively, why don't you go to the people and work for revolution? [10] Ting rejected the populist argument, just as he rejected the argument that social reform should take precedence over political reform, and for the same reason. Bad government was far more the fault of the leadership than of the people led, and reform must begin among those with political responsibility and the power to have immediate influence. He could sympathize with arguments that the group become an active political party, for it was his own ambition. He did not, like Hu Shih, suggest that an overriding need for unity ruled out the formation of a more conventional political association. A party simply was too expensive, Ting said. From the start, party politics under the republic had been undermined by financial weakness, which had forced all parties, both administration and opposition, to depend upon the government in power for subsidies. It was foolish for the *Endeavor* group even to consider party action until their treasury, and therefore their votes, could be utterly free of all outside dependencies. In the meantime the group did consider itself a kind of protopolitical band, dedicated to preparing itself for a future role.[11]

For Ting, however, political preparation was first of all a kind of moral training. He envisioned the small group of *Endeavor*'s creators, perhaps a half dozen enthusiasts, meeting regularly to establish the core of a "good government party." Their first duty was the cultivation of their personal integrity: "One: we must preserve our qualifications as 'good men' — that is to say, we will not 'do that which is unbeneficial.' Positively, we will practice self-discipline in our personal conduct and do ourselves what we as critics demand that others do . . . Two: we must be professional men, and, moreover, we must enhance our professional capabilities. Three: we must find ways to keep our standard of living from continuing to rise." [12]

The next step in political action is to make judgments concerning others who are potential allies. The "good men" should recognize "that there are powerful men in society and politics, and calmly and dispassionately investigate their character, temperament, and capabilities, in order to determine what our attitude toward them should be." Here Ting added a rubric which was a constant theme of his discourse with fellow intellectuals: "Military men are included among those of whom I speak, because military men are also citizens of the nation, and there are also good men among them. It is not proper to use common abstract expressions like 'militarist' and 'warlord' in speaking of individuals." [13]

However, concerning revolution itself Ting shared *Endeavor*'s addiction to militant rhetoric, together with its readiness, when seriously challenged, to admit that a policy of violence was not a possible choice. The image of popular strength aroused by the May Fourth demonstrations remained dazzling. Hu Shih could prophesy that the people would use strikes and boycotts to force a north-south peace conference to do their will, and he even threatened the government by raising the specter of "the method of guns and bombs" (terrorism).[14] All *Endeavor* editorialists, including Ting, warned that protracted obstruction by the authorities would end in violence and in a revolutionary new mandate. "If a great political transformation happens to occur, we must demonstrate a spirit of sacrifice, an attitude of total commitment, and we must struggle even to doing battle," Ting wrote.[15] But when asked directly whether the *Endeavor* group favored a policy of armed violence, Hu Shih responded with riddles: "At times construction is destruction; at times destruction is construction." [16] Ting was more frankly pessimistic. He admitted that open revolution, openly fought, had been successful at times in Europe, but he noted simply that "it requires an army." The only violent actions he imagined as being within the power of educated Chinese were individual acts of terrorism — which he repudiated as both morally detestable and unprofitable.[17]

Ting's protopolitical band really amounted to no more than another study group, dismally scanning the horizon for allies among the real power structure. Their bid for influence rested in the hands of the Chihli clique. *Endeavor* in fact relied upon the rump "Old Parliament" in

Peking as the machinery for guiding Wu P'ei-fu into constitutional channels. A basic assumption behind their proposals had been, in Ting's words, "to assume good intentions in the men of today." [18] As a token of their good faith, three members of *Endeavor*'s inner circle accepted positions in the first cabinet formed under the restored president, Li Yüan-hung, which took office on September 19, 1922. Wang Ch'ung-hui became premier, Lo Wen-kan served as finance minister, and T'ang Erh-ho became minister of education. Their association with the *Endeavor* manifesto brought upon the heads of the new ministry the half ironical sobriquet "the good men cabinet."

Even before the new cabinet took office, *Endeavor* had written of the reassembled "Old Parliament" in severely admonitory terms, recalling its past venality and pointing out that rumors of present payoffs were already circulating in Tientsin. In fact the assembly was corrupted from the start by the intrigues of Wu P'ei-fu's nominal lieutenant, Ts'ao K'un, to replace Li Yüan-hung as president of the republic. Ts'ao's influence was enough to keep the new cabinet from receiving any parliamentary approval, and within three weeks of their assumption of office, the "good men" were already threatening to resign. Wu P'ei-fu's frail allegiance to constitutionalism could not survive the threat of a serious quarrel with his army officers, and Ts'ao K'un's campaign was allowed to run its course. [19]

By October, *Endeavor* editorialists were writing that the acts of parliament were "like a bucket of cold water" on the group's hopes. The blow fell in November. On the morning of the 18th Lo Wen-kan was placed under arrest by order of President Li, charged with accepting a bribe in connection with a treaty with Austria. This accusation was made in a spirit of cynicism so profound that it was reported President Li himself consoled Lo for his detention, pointing out that such persecution would only enhance his reputation for integrity! [20] All that was left for *Endeavor* was to ask the question "Where is the law of the land?" and to print the unhappy answer: "From the President above to the small officials below, men find the law a nuisance and execute orders to their liking. How can the nation have lawful government, and how can the people's rights be preserved?" [21]

As for action, only one course seemed both open and appropriate — the time-honored one of resignation. The cabinet itself immediately resigned in protest, and they were followed a few weeks later by the dean of the *Endeavor* group, Ts'ai Yüan-p'ei, who stepped down from his post as chancellor of Peking University. In a somber statement Ts'ai declared that the Lo Wen-kan affair, and the subsequent appointment of P'eng Yün-i as minister of education, had convinced him that he could no longer protect the freedom of his staff and colleagues. He went on to reflect that he had often considered the idea of political protest by means of mass resignation of the educated and trained bureaucrats, since without the services of trained talent it seemed to him unlikely that any administration could carry on. However, he added, "I have refrained from promoting this idea because a course of action which is easy for me may be difficult for others." [22] Ts'ai was right. His veiled appeal for a protest from the intelligentsia by means of mass withdrawal produced few ripples in Peking. Only his close associates Shao P'iao-p'ing and Chiang Meng-lin followed him into retirement.[23] The best that could be said — and Hu Shih was quick to say it — was that Ts'ai's personal example redeemed the honor of the educated class.[24] A bitter epigraph of the whole experiment was supplied by the ex-minister T'ang Erh-ho, who later said to *Endeavor* in the person of Hu Shih: "May

I urge you not to talk politics. Before, when I read your articles, I felt that they were not without some good sense. After I entered the government, I saw that as a matter of fact nothing was like that. Hardly a sentence of what you said scratched where it itched. What you said was one world; where I travelled was another." [25]

The good government experiment had been a total failure. In fact, it was the last important effort by independent intellectuals to cooperate with a major warlord government. Only after the final success of the Kuomintang in 1928 did the "best elements" find another military ruler capable of rallying their patriotic energies. Briefly reinvigorated by the excitement of the science and metaphysics debate, *Endeavor* returned to the intellectuals' alternative mission of educating the Chinese for modernity. But its militant spirit was broken, and in November 1923, when the election by bribery of Ts'ao K'un as president of the republic demolished the remaining scraps of parliament's reputation, the magazine also succumbed.

Throughout its brief history the *Endeavor* movement had been handicapped by a mistaken estimate of warlord power on the one hand and a faulty analysis of the politics of mass movements on the other. The group's miscalculations were fostered in turn by some of the weaknesses of pragmatism as a political concept. Hu Shih's popular pragmatism did not represent a systematic social philosophy. In his rather free adaptation of Dewey's doctrine of scientific "experimentalism," Hu described the experimental method simply as one which focuses attention upon the concrete problems of life and teaches human beings to deal with them by the quasi-biological process of response and adaptation to changing environmental conditions.[26] By suggesting that the practice of "experimentalism" involves the conscious shaping of human conduct to conform with the evolutionary forces of change and that the scope of this method is as broad as the human enterprise itself, Hu was able to make the doctrine of pragmatism into a philosophical justification for the new culture movement. The experimental method was also a revolutionary method, according to which the analysis of concrete difficulties would enable the Chinese to devise new and creative solutions to their contemporary problems. By calling upon men to react to the real world of modern challenges rather than retreat to the increasingly irrelevant solutions supplied by traditional precedent, pragmatism was in fact a progressive force. But its purely methodological focus — and the way which the methodology was understood to operate — kept pragmatism from serving as a really revolutionary doctrine.

Ting and Hu believed that in essence pragmatism consisted in scientific reasoning applied to practical problems. Modeling their theory on their own particular interpretations of the nature of scientific thought, they assumed that concrete, empirical examination of evidence concerning a given problem — political or otherwise — was bound to yield the one, proper, scientific solution to it. One built up a program of political action from classification and testing of the concrete data of political experience. One did not, they thought, reason from a priori doctrines about the nature of man and government. However, this view of government is far more successfully assimilated by a theoretical conservative, who works within the security of an established political tradition, than it is useful for a theoretical reformer, who faces a need for drastic innovations. Political acts must carry with them assumptions about human nature and goals; although the conservative can accept the practices and values of his tradition as part of the empirical "given" of his analysis, and still hope for success in action, the radical reformer must be an ideologue and invent the justification for his changes. The *Endeavor* pragmatists found it possible to dismiss the battle of *isms* derived from Western political theories as pedantic and irrelevant, but when they considered their own political program, the demand that they examine the "data" of Chinese experience made them introduce as part of the unconscious content of that "data" many traditional political patterns. Their idea of the intellectuals' relation toward power on the one hand and toward the people on the other as well as their notion of the function of bureaucracy and of the links between ethics and government were all telling illustrations of the fact that certain Confucian political structures still seemed to them among the given facts of any political situation. For all its progressive aura, a loosely applied notion of scientific method in politics had a subtly conservative pull. Without realizing it, by rejecting ideology and putting their faith in the scientific method, the *Endeavor* pragmatists left to others the speculative freedom to be genuinely radical.

In a context where the republic itself was a post-Confucian innovation, this conservatism did not reveal itself in the pragmatists' actual programs, but on the more abstract level of their beliefs about the structure of the political process itself. Continuity with the past lay in certain of their categories of political analysis — categories which did not extend to specific proposals, but which had a powerful unconscious influence on the kind of specific proposals they were likely to consider.[27] For example, the *Endeavor* platform was conditioned by a Confucian notion of political relevance. On the one hand, editorialists carefully devised purely administrative solutions, ignoring the fact that no political environment existed in which these solutions could be implemented. *Endeavor* abounded in articles outlining constitutional formulas, budget proposals, and blueprints for some imaginary disarmament conference. Or, in a spirit of generality as excessive as their administrative plans were minute, its contributors wrote of the fundamental evils of war, anarchy, and corruption, and of the horror these things inspired in all men of good will. Their rhetoric expressed little more than a common abhorrence of sin and the hope that it could be stamped out by the effort of will invoked in the journal's title. These were the "concrete evils" and the "practical solutions" of their pragmatic theory, but in true Confucian style they were operating according to a view of government which alternated between a preoccupation with pure administration and a concern with pure ethics.

Then too the *Endeavor* group's reluctance to devise a revolutionary program was in part a consequence of their emphasis on the leadership of an educated elite. As a whole, they lacked any real contact with the masses as a source of political power. May Fourth remained an example which they could not imitate except in rhetoric. Their plans for political action took place during domestic evenings among intimates — not in the sweaty commerce of meeting halls and rallies. Their attempts to win influence focused upon the cabinet or bureau, and when they thought of power, they imagined officials or generals, not militias and mobs. None could escape the contradictions of a reform program which would have to depend upon popular power, if it was to have any weapons at all, and a leadership which had no instinctive rapport with that power.

In spite of their acquaintance with Western democracy and their talk of forming a Western-style political party, the *Endeavor* leaders were too much steeped in the traditions of an intellectual aristocracy to contemplate political leadership except as acting upon a passive body politic. They accepted the idea of parliamentary government, but not the ideals of democracy. Where Westerners have tended to think of a parliament at times almost mystically, as the expression of the will of the community, and always as a body in intimate interaction with its constituency, the Chinese liberals saw it as an apparatus for selecting leaders. Their view of democracy was, in Hu Shih's favorite pragmatic idiom, "instrumental." It offered a technique for the orderly selection of the best men for positions of power and for the transfer of power from one set of men to their successors. The fact that mass illiteracy made it seem unwise for universal suffrage to be attempted for many years in China helped to reinforce their feeling that democracy merely supplied new machinery through which the "best elements" would play a familiar role. In describing the ignorance and unreadiness of the people, they saw what was true: the uneducated peasant was indeed not capable of assuming political leadership. But neither did they think of him as a political participant of any importance.

In a nation suffering from civil war it is easy to point to the futility of political ambition in a group of unarmed intellectuals. But here Ting and his associates were perhaps less blind to their weakness than alive to a felt responsibility which did not permit calculations of the odds. Something compelled them to have such ambitions, to feel that political aspirations were their duty no matter what the difficulties, and moreover, to believe that they were owed some kind of recognition by the rest of society. For centuries China had been administered by a Confucian bureaucracy that was trained to cultivate moral authority, intellectual achievement, and political leadership as part of a single ethos. The literati were the bearers of the social and ethical ideals of the community, and their theoretical position vis-à-vis the imperial power was in part that of king's factotum and executor and in part that of his tutor and judge. Confucian statecraft had no easy recipe for overcoming the obvious dangers and contradictions of such a role. It had only a concept of social duty and the theory of the efficacy of example, whereby the upright official was taught a stern notion of his responsibilities and led to hope that his personal conduct would have an influence in molding the conduct of others.

One of the strongest characteristics of the *Endeavor* group was this peculiarly Confucian belief in the social responsibility and prestige of the educated "best elements." It was demonstrated not only by their own words, but also by the chorus of approval which greeted their manifesto when it first appeared.[28] Such assumptions about government also made it possible to hope, in the face of Republican experience, that their initiatives might be successful. The theme of good government and of good men itself suggested this. In spite of a Deweyite cast, these slogans were also congenial with the idea that qualities of character and scholarship are the proper and effective basis of political authority, capable of filtering down from the leadership to have a purifying effect upon the entire body politic. The requirement of technical skill was the group's modern addition to the traditional standards for statesmanship, not a substitute for them. They did not imagine that a competent but personally lacking individual could have the right to authority. Secondly, they shared an administrative view of government: they assumed that this was the level at which most important political decisions were made, regardless of the locus of formal sovereignty. Again this was a view natural to a Confucian bureaucrat who participated in ruling a vast area by means of a special corps of officials in a manner which saw rather few basic changes of policy over long periods of time. It made the *Endeavor* group willing to claim that the failure of the Republican experiment was due to the refusal of good men to stay in the administration. At times they even talked as if the military were simply another branch of the bureaucracy. They were more than merely indignant that educated men were being governed by illiterates, but also believed that the military themselves must recognize the propriety of intellectual leadership. Ting had a keener instinct for military power than some, but it led him to an image of the general-as-bureaucrat. He suggested that innate worth and special education might also qualify some militarists for the ranks of "good men." So the gap between physical might and moral authority might be bridged — in the person of an individual.

Ironically the *Endeavor* participants were impelled toward political action by these traditional factors, but at the same time they had rejected many of the psychological attitudes which helped the old Confucian official accommodate himself to the limitations of his influence. Without the panoply of empire to sustain it, "loyalty" was no longer an administrative virtue; the abstraction, the nation, was replacing the more manageable family, clan, or district in their minds as the proper beneficiary of their political labors; the negative concept of government, which permitted imperial officials to content themselves with customary rounds, had been rejected in favor of a program of modernization that drastically increased government's scope. Where the mandarin's comfortable careerism was a vanishing possibility, intellectuals continued to feel the imperatives of the Confucian official ideal.

After *Endeavor* ceased publication, most of its supporters settled down to endure the warlord regime stoically from the sidelines, though a few turned to the openly revolutionary alternative of the Kuomintang-Communist axis. The man seemingly least affected by the lessons of recent experience was Ting Wen-chiang. Scorning withdrawal, unable to embrace revolution, he carried on by acting as if the collapse of the good government experiment in no way invalidated the assumptions upon which it had been built. In one of its last issues *Endeavor* printed a speech Ting had recently made to the students of Yenching University. Entitled "The Responsibility of a Few Men," it was a plea for the continued political commitment of the educated classes. It reiterated the claim, implicit in *Endeavor*'s apotheosis of "good men," that history is made by acts of individual greatness, and it declared progress the servant of inspired men who by will alone force circumstances to yield to them.

"The disorder in our Chinese politics is not because the citizens are still at an immature stage of progress, nor is it because of the corruption of officials and politicians, nor is it because militarists and warlords have abused their arbitrary power — it is because 'a few men' lack both the sense of responsibility and the capacity to bear responsibility." [29] Calls for heroic Confucian leadership have long echoed through the writings of scholars facing the many crises of the imperial monarchy. In repeating this call, Ting explained it in terms of Darwinian biology, which taught the native inequality of man. Nature bred superior men, the "few," and in society these few always controlled events. To his democratically inclined student listeners he was ready with reminders that organized administration is the core of all governments, democratic or autocratic, and that "organized government obviously is the

161

affair of a few men." Democracy, he said, is merely an experimental form of rule designed to correct the defects of monarchical forms by making certain that the leadership is always drawn from the worthy, and to enable the majority, within the framework of being governed by the few, to keep their rulers under surveillance.

But in fact democratic ideology was an obstacle to progress in China because of the very vastness of the social change it presupposed. Its adherents imagined that democracy would flourish when the populace had been educated to practice it, and they cultivated the garden of social and educational reform, while remaining insensitive to their immediate political responsibilities as an elite. For Ting, society almost seemed divisible into specialized castes, identified by the performance of distinct functions. The masses are farmers and laborers, who, he pointed out, are less unprepared for their appropriate duties in life than are the members of the governing class. "It is obvious," he said, "that China has been brought to its present position by presidents who do not know how to be president, prime ministers who are unfit to be prime minister, deputies who are inadequate as deputies."

In Ting's mind the solution to the political crisis lay in a substitution of personnel. "Soldiers and bandits are resourceful enough to make themselves warlords," he observed bitterly. "Why aren't educated and patriotic men resourceful enough to do the same?" The failure can only be attributed to a lack of spirit and resolution on their part. Ting looked for its cause in three hundred years of Ch'ing despotism, which corrupted and enervated the ruling classes and schooled virtuous men in cynicism and apathy.

Throughout the lecture Ting wavered uneasily between his instinct as a scientist to assign material causes to the phenomena of history and the idealist rhetoric of patriotic feeling. Environment and heredity gave this double perspective to his hero. Ting stated that a good man must be born in an environment which allows his qualities an opportunity to develop and that a leader is a man who gathers together prevailing popular sentiment and links it with opportunities for action; yet he also declared that the hero's inborn character and intelligence should make possible upthrusts of daring and resolution which lift the individual to a mastery of fate. In this way the past Ch'ing bureaucrats were imagined to have been unmanned by a corrupt environment: "One who did not dare venture forward was called high-minded; one who did not discriminate right and wrong was called loyal; one who did not know practical world conditions was called refined; one who did not care to be trained in affairs was called wise; one who abhorred hard work was called a *bon vivant*." [30] It was even predicted that the young Chinese of the future, purified by modern education, would live in a favorable environment which would naturally and easily foster political enlightenment. But by contrast the present-day audience could rely on no external supports and was exhorted instead to a superhuman effort of will: "China today need not fear diplomatic failures; she need not fear the bankruptcy of the Peking government; she need not fear north-south war. Most to be feared is that knowledgeable and virtuous men are unwilling to make political endeavors . . . we should not rely on any man . . . we should not fear any man . . . we must be willing to unite and submit to discipline . . . Only eighty thousand Chinese are college students who know a little about science and have read a few foreign books. If we are not the tiny elite, who is? If we do not have a sense of responsibility, who will?" [31]

Ting offered his listeners two examples of personal heroism, one from the West and from the East. Significantly, the Western model was revolutionary — the Sinn Fein rebels, whose spirited fight for Irish independence had aroused Ting's admiration during his stay in England. But at home in China he chose Tseng Kuo-fan, a man of action who symbolized conservation, not renovation, stability, not upheaval. Ting disassociated his praise from the imperial system Tseng served so effectively: "It is his determination and ability to act which we honor." But in fact he was responding to a whole series of Confucian images: the man of gentry birth and education, the civilian self-taught strategist who made himself the match of generals, the fighter for an orderly, benevolent, and paternal administration against both a superstitious rabble on the one hand and a corrupt, temporizing officialdom on the other. The "good man" of Ting's patriotic fantasy was a scholar-general.

In the wreckage of the *Endeavor* experiment Ting attempted to carry on by a voluntaristic appeal for heroes, not by a call for new institutions or for a new pattern of social relations among the people of China. For the moment he seemed even to have lost his confidence in the progressive power of economic industrialization. He had reached the hard, practical conclusion that reforms, both political and economic, were at the mercy of the militarists; he was prevented by his concept of the intellectual's status from seeking a solution, as the revolutionaries did, in a new alignment of popular forces. His serious interest in republicanism did not survive the illusion that it would provide an up-to-date structure wherein an administrator descended from the Confucian tradition could exercise his proper influence. Instead, he began to dream of a military hero who might synthesize modern and traditional values:

one who would govern with paternalistic concern for the people's welfare; one who was educated in modern learning, since war and government today require Western techniques; and one who could command the coercive power which is the true basis of all political authority. Such a hero, embodiment of a triumphant will, should become the new model for patriotic youth.

Chapter VII

Director-General
of Greater Shanghai

In the late autumn of 1925 Ting formally resigned his post with Pei-p'iao mines, ostensibly in order to accept a temporary assignment on the Sino-British Commission for the redemption of the British portion of the Boxer indemnity. However, his work on the commission actually was only a prelude to a second plunge into active politics. From May through December 1926, Ting held the post of director-general of the Port of Greater Shanghai (*Sung-hu shang-pu tsung-pan*) under the auspices of Sun Ch'uan-fang, warlord of five lower Yangtze provinces. In the eyes of the Kuomintang, Ting was on the side of the enemies of the Northern Expedition. In the eyes of many of his independent intellectual friends of *Endeavor* he was supporting one of the militarists whose factionalism had been responsible for the civil war and the collapse of constitutional government. As for the revolutionary left, to them he was another of the Westernized Chinese products of the treaty ports and a foreign education who served as lackeys to warlords and imperialists. The episode won him wide unpopularity and clouded his reputation and even his career for some years afterward.

In fact, however, it was merely an attempt to carry out on a local scale the principles which had been enunciated for a national audience in "The Responsibility of a Few

Men." The condition of the Peking government had provoked Ting to a true diagnosis and a false conclusion. The diagnosis was that power in China lay with the competing militarists, and that for the time being authority could be exercised effectively only on a regional basis. The conclusion was that the formula which had failed with the Peking government might succeed on the more manageable scale of the provinces, given the fortuitous emergence of "good men" among the military leadership. The way to make military power constructive was to domesticate it with a surrounding civilian bureaucracy — to provide each general with a scholarly entourage to make his intelligence "educated." Such a course seemed possible to Ting because basically he thought that "patriotism" in military men was equivalent to their accepting civilian goals.

Ting had thought about the problem of militarism often before. His first attempt to domesticate the armed forces had consisted of a typically omnivorous attempt to assimilate information. In the early 1920s he gradually became exceptionally knowledgeable about military conditions in north China. He cultivated acquaintances among army officers and encouraged distant friends to send him reports of their observations of local garrisons. His own frequent travels in Manchuria on behalf of Pei-p'iao mines gave him an opportunity for especially close study of the Fengtien and Chihli forces. *Endeavor* published a series of articles on the nation's many armed forces, to which Ting contributed pieces on military conditions in Manchuria, Hunan, and Kwangtung, together with an analysis of the first Fengtien-Chihli war.[1] An expanded version of these articles was later published as a book, *Notes on Republican Military Affairs,*[2] which attempted a brief history of the provincial armies of north China as well as an analysis of the structure, location, size, and preparedness of each as of 1925.

"Complete disarmament, not just reduction of armies!" was one of Ting's personal contributions to the slogans of the *Endeavor* manifesto.[3] In *Endeavor*, Ting asked that concrete studies of the disarmament problem replace rhetorical denunciations of warlordism, and he praised a book by his friend Chiang Pai-li as one of the few serious attempts by a knowledgeable military man to study the subject.[4] All of this careful examination was typical of Ting: scientific analysis of armies was offered in the hope that it might lead to a more pacific military policy.

Disarmament was of course illusory, but Ting's hope of taming the military through education persisted. His acquaintances among army officers extended to the highest ranks and included men like Chiang Pai-li, Yen Hsi-shan, and Chang Hsüeh-liang. To his civilian intellectual friends he preached against literati prejudice: many Chinese officers were men of natural talent and sincere love of country. The fault lay in their environment, which left them deficient in the modern education needed to fit them for the serious responsibilities of national leadership. "Within the limits permitted by their education and environment, they also wish to save the nation and contribute to society." [5] Luckier parts of the world already enjoyed the benefits of a military intelligentsia, he believed. One of Ting's acquaintances in Peking was the retired American general William C. Crozier, a West Point graduate and a former wartime aide to President Wilson; this man seemed to embody a distant ideal of political and educational sophistication in a general, and less explicitly, of deference to civilian values.

In the meantime Chinese military education appeared to Ting to be "the worst in the world." He deplored the fact that native army officers learned nothing of modern history, politics, or economics; even this deficiency was less shocking than the sloppy quality of their instruction in the essentials of military science. His own technical expertise as a geologist and as a businessman gave him a standard for personally measuring military deficiencies in cartography, strategy, and supply. He noted that instruction was carried out by graduates of inferior Japanese military schools through translators, and that students used written translations of foreign texts — all of this years after civilian universities had abandoned such practices. A logical sequence to the belief that army officers need education is the desire to educate them: if scholar-generals cannot be found, they must be created. Ting began to nourish a private ambition to organize a modern military academy. In the autumn of 1925, when he first discussed with Sun Ch'uan-fang the possibility of entering the latter's employ, he made the astonishing proposal that he be placed in charge of the general's officer-training schools. Sun laughed. Later, reporting the incident to a friend, Ting laughed too, but in a different way.[6] In spite of this wry indication that he was aware he may have appeared a fool, ten years later, on the eve of his death, Ting said privately that to undertake the reform of military education was still one of his cherished ambitions.[7]

Unable to make the Chinese officers' corps over in his own image, Ting seriously overestimated the satisfactoriness of the model already in existence. In explaining his confidence in military leadership, he relied heavily upon the concept of "patriotism," which evoked in a Western-educated intellectual images far different from those stimulated in a professional soldier. Around a common nucleus of hatred of foreign oppression clustered diverging colonies of meaning. Ting's personal patriotism was both nationalistic and civilian: his knowledge of the outside world made him see China as a distinct, even parochial, unit, whose national identity would be affirmed by the acceptance of a centralized modern government. But he accepted as "patriotism" in a militarist a set of purely negative characteristics. The warlord who limited himself to a defensive policy of peace in his territory, kept his troops from tyrannizing the local population, avoided forced loans and surtaxes, was restrained in graft, and shunned foreign entanglements, had by these parodies of civilian political virtues earned Ting's praise. For example, he accepted the legend of Yen Hsi-shan's "model province," admiring its strong local militia and police and praising Yen's emphasis on education and organizations for moral propaganda.[8] However, in fact Yen kept local peace by means of a series of shifting alliances which perpetuated the worst instabilities of the warlord system; his allegiance to Chinese unity always evaporated on the day when its prospect seemed to threaten his own autonomy.[9] As a gentry-bred civilian, Ting assumed that a military governor's administrative apparatus was designed to further purely political progress rather than to support a military occupation. As a modern nationalist, he assumed that love of country necessarily wore a nationalistic dress, and he put his faith in a military patriotism which in fact turned out to be merely antiforeign and which found expression in personal rule. With these illusions Ting turned to Sun Ch'uan-fang in a moment of immediate danger to his native province of Kiangsu.

After the fall of the "good government" cabinet, Wu P'ei-fu and Ts'ao K'un had not controlled Peking for long. In 1924 new fighting broke out between their forces and the Fengtien faction and ended with the collapse of Ts'ao K'un's administration. Beginning in November of that year the capital was ruled by the so-called "Executive Government," a loose coalition of Wu P'ei-fu's enemies, among whom Chang Tso-lin enjoyed military preponderance and dominant influence. During this period of uncertain conflict in the north, the central coastal provinces suffered from extreme instability. In September and October of 1924 and again in January 1925, Kiangsu was involved for the first time since the 1911 revolution in a local military campaign. Its opponent was Chekiang, and the issue was a long-standing contest for the control of the city of Shanghai and its rich revenues. In *Notes on Republican Military Affairs,* Ting counted the cost of this small war to Kiangsu: a swollen army of recruits to support, an 8,000,000 yuan debt from a loan raised to finance the campaign, several ruined cities, the interruption of commerce, and universal forced exactions.[10] One consequence of this inconclusive campaign was that Wu P'ei-fu, having been driven from the capital, now found his authority eroded in the Chekiang-Kiangsu region as well. To fill the subsequent vacuum the most likely candidate proved to be his nominal lieutenant, Sun Ch'uan-fang, who earlier had been Wu's military governor of Fukien. Sun had taken advantage of the Kiangsu-Chekiang conflict to move northward rapidly from his base in Fukien, and as a result of the war he was able to establish himself as military governor of Chekiang too. In the course of 1925 he consolidated his position as the military ruler of the five provinces of Chekiang, Kiangsu, Anhui, Kiangsi, and Fukien.

At the same time Chang Tso-lin's Fengtien clique was also anxious to expand into the central provinces. Shanghai had already endured a few weeks of occupation under the Shantung army of Chang Tso-lin's ally Chang Tsung-chang, who had moved south briefly in January 1925 in the name of "restoring order" to Kiangsu and Chekiang.

It was feared that he would soon return as the vanguard of Chang Tso-lin's drive for nationwide domination. To the Chinese gentry of Kiangsu, as to most educated Chinese, Chang Tso-lin and his associates enjoyed the most evil of reputations: for disorderly troops, racketeering, extortion, and underhanded dealings with the Japanese. Ting's experiences in attempting to deal with the Manchurian authorities on behalf of the Pei-p'iao mine had left him with no illusions concerning the detestableness of the Fengtien faction.[11]

These alarms had prompted the formation in 1924 of a clandestine society to "save Kiangsu," in which Ting was active together with his friends and fellow provincials Liu Hou-sheng and Ch'en T'ao-i. Ch'en I of Chekiang was also associated with the group.[12] In August or September of 1925 Ting received a secret telegram from Lo Wen-kan suggesting that he go to Yochow to see Wu P'ei-fu,[13] who conceivably might still be capable of rallying an anti-Fengtien coalition. Ting did so, but the meeting apparently was inconclusive. However, when passing through Shanghai, Ting found the members of the society to "save Kiangsu" discussing ways and means for raising an army for the province's defense, and in Hangchow he found Sun Ch'uan-fang actually in the process of assembling such a force. The conclusion was obvious. Ting spent a week at Sun's headquarters and two months later he had resigned his post with Pei-p'iao mines.[14]

Sun's troops easily repulsed a tentative invasion by Chang Tsung-chang, and in May 1926 Ting and Ch'en T'ao-i were established respectively as director-general of Greater Shanghai and civil governor of Kiangsu under Sun's avowedly independent Federation of Five Provinces. Ting's old associate Chiang Pai-li, who had been offered the job of director-general but had declined, also joined

Sun's government as an informal military adviser. The Kiangsu gentry had thrown in their lot with Sun at about the same time that the International Concessions at Shanghai did so, and for much the same reasons. However, in Ting's case the desire for peace and commercial self-preservation was less important than his declared principles in "The Responsibility of a Few Men." Predisposed toward paternalistic rule, convinced of the obligation to act, disillusioned by the lack of a central framework, his mind focused on the administrative solutions possible under a stable local ruler who would allow initiative to educated experts. Sun Ch'uan-fang qualified as a "good" warlord in his concern for order, discipline of troops, restraint in taxation, and his declarations of patriotic dedication. For a time Ting even hoped that Sun's armies might be capable of bringing about military unification.[15]

At the same time, a second powerful factor impelling Ting toward an official post was the nature of the job he was offered in Shanghai. The conditions of extraterritoriality in China's fastest-growing port and industrial metropolis had made Shanghai into an acute center of Sino-foreign conflict. Ting's assumption of office in May 1926 was also a patriotic response to the May Thirtieth incident of 1925.

In mid-1925 protests against conditions in Shanghai's foreign-owned factories and sweatshops had led to an explosive confrontation. Indignant over a recent strike incident, demonstrators gathered in front of the International Settlement police station on May 30, and a number of them — mostly students — were fired upon and killed by Settlement police. Three weeks of demonstrations followed, which pushed the city to the brink of anarchy. Sympathy strikes in most other major cities gave the movement some of the dramatic fervor of May Fourth, and it soon ranged far beyond a labor protest to one against the entire range of imperialist privileges. But these demonstrations had an organizational coherence and an ideological thrust absent from May Fourth. The Kuomintang and Communist parties were by now active in Shanghai, and Communists dominated the Shanghai General Labor Union, which acted as an umbrella for most groups active in the strikes and demonstrations. Their aim was to create a mass political movement, and in so doing they supplied the public with an interpretation of the unequal treaties as the imperialist instruments of an unbridled capitalism. The May Thirtieth movement accentuated the drift leftward in Shanghai and other large Chinese cities and vastly increased the radical parties' capacity for revolutionary political leadership.

For the ordinary citizens of Shanghai, the unequal treaties were directly linked with long-standing inequities of daily life under a foreign-dominated municipal government. As a result of the movement there were renewed and more embittered local demands for such things as Chinese representation on the International Settlement's Municipal Council and on the council's Police Commission, freedom of speech and press for Chinese living within the Settlement, immediate rendition to Chinese jurisdiction of the foreign-controlled Mixed Court, and an end to Settlement attempts to build and police roads outside its boundaries.[16] The Peking government, however, was too weak to force either a general reconsideration of the treaty system or any redress of grievances in Shanghai. On the national diplomatic level the May Thirtieth movement led to nothing. Inside Shanghai it left a mood of resentment which threatened to undermine the day-to-day functioning of administration. Fearing this, and frightened by recent experiences of police and mob violence, the

Shanghai foreign community wanted to evade the larger diplomatic issue, and yet to restore local peace. At the end of the violent summer of 1925, British leaders issued a plea for calm and declared themselves in favor of steps "to give effect to the declarations of the Washington Conference." [17] Soon thereafter it was announced that negotiations would resume on the rendition of the Mixed Court, and the Shanghai Municipal Council voted to admit three Chinese to its governing body of twelve. It was clear that some possibilities for compromise on concrete issues lay in the hands of local officials, Chinese and European, if they confined themselves to the issue of the interpretation of the treaties, not of their existence.

From the beginning of the May Thirtieth incident Ting had been prominent in protests designed primarily to convince the British of the legitimacy of Chinese grievances. He was a signer of the "Tientsin Manifesto," issued on June 7, 1925, by Liang Ch'i-ch'ao, Wellington Koo, and other moderates. Written as a proposal for a negotiated settlement, this open letter called for calm on both sides, insisting nevertheless that the burden of responsibility for violence lay with the British Settlement police. It ended with the following warning to Westerners:

Times in China have greatly changed in the last decade, and though education is still insufficient as measured by the actual needs, the general standard of enlightenment has risen considerably . . . So much of the welfare of the foreign community in China and the interests of foreign trade with her depend upon the existence of a feeling of mutual understanding and confidence between the Chinese and foreigners in China, that, if not on the high ground of justice, at least as a matter of sheer expediency, it would be well for the foreign nations or their agents in China . . . to seek to understand the viewpoint of the Chinese people and at least in some measure to consult their interests in matters vitally affecting them.

However, it also cautioned the Chinese against disorderly gestures against the treaty system as such.

No class of the Chinese people should seek to antagonize the foreign residents of China or the countries they represent *merely because they still hold on to their special problem of treaty revision* [italics mine], and readjustment of the basic relations between China and the foreign nations cannot be settled . . . by coercion, force or violence on either side.[18]

Another letter, composed by Ting in English and signed by himself, Hu Shih, Lo Wen-kan, and Yen Jen-kuang, was fortunate enough to reach the attention of the Trades Union Congress in London. Under the title "China's Case" it was reprinted and circulated widely, and it formed one of the documents upon which the Labour party based its parliamentary campaign against the Foreign Office's handling of the incident.[19]

By his part in these initiatives Ting showed he believed that the most effective method for recovering Chinese sovereignty was through negotiation. Perhaps more important, it was clear he had accepted the British values behind the rationale that the treaty system had been a necessity imposed by Chinese backwardness. To defend their threatened prerogatives, Europeans relied heavily upon the argument that changes in Shanghai's international status were impossible, given the present inadequacies of Chinese municipal rule. The legations reminded Peking's foreign ministry that Chinese reform was a prerequisite to any modification of the system of extraterritoriality. The Foreign Committee Investigating the May Thirtieth Affair emphasized the need for "a responsible Chinese authority to keep firm control over the Chinese districts of Shanghai." [20] Ting himself was too committed to the administrative and economic Westernization of China to see these British objections as other than reasonable. Faith in reason led to confidence in British good faith. Since the ground for dispute rested upon the objective fact of Chinese deficiencies, a change in objective conditions should bring a peaceful end to Western privileges.

Although some might question the practicability of such long-range reliance upon the resources of reason in international conflicts, Ting's determination to devote his talents to the short-term improvement of conditions in Shanghai was consistent with his idea of political pragmatism. His policy promised more immediate benefits to the people of the city than did the doctrinaire posture of the Kuomintang and Communist party radicals, who wanted to exploit the "Shanghai case" as a propaganda issue in their overall revolutionary strategy. The radicals claimed that the introduction of a modern administration in a Chinese city was no legitimate concern of any foreign power; but to Ting this meant that Shanghai would be left without a policy until a revolution could make one possible. Ting and his fellow pragmatists were unable to appreciate what the revolutionary left did not forget — that the instrument which brought Europeans to the bargaining table ultimately had been, and would continue to be, the coercive pressure of the Chinese masses.[21]

Ting was formally offered the post of director-general of the city of Greater Shanghai in April 1926. At the time he was visiting the city as a member of the Sino-British Commission for the rendition of the British portion of the Boxer indemnity.[22] Ting's acceptance was smoothed by signs that the British were willing to consider the goals of his administration favorably. Prominent among those who encouraged him was Viscount Willingdon, chairman of the commission's subcommittee in China and a former viceroy of India.[23] Ting's inauguration, on May 5, 1926, was greeted with approval by the *North China Herald*, the China Association, and other organs of British opinion in the Settlement. He remained in the post for eight months, until December 1926, and during that time also acted as Sun Ch'uan-fang's informal spokesman in foreign affairs.

Diplomatic negotiations, then, took priority in Ting's administration, and in fact his most enduring accomplishment turned out to be the reform of the International Settlement's Mixed Court. By treaty this was supposed to be a Chinese court which allowed foreign consuls to oversee cases involving treaty-power nationals. But in the years since the Treaty of Tientsin and the Chefoo Convention, it gradually had become a foreign-controlled court of first and last instance for all residents of the International Settlement, including the Chinese. Procedural abuses were commonplace as well: the court did not honor rulings of Chinese courts elsewhere; it provided no machinery for appeal; and only foreign lawyers were allowed to appear before it. For years the Chinese authorities had been complaining about these injustices, and the May Thirtieth incident had created such an acute mood of general resentment that effective functioning of the court itself was beginning to be impaired.[24]

Four months of direct negotiations led by Ting and a team of local Chinese jurists resulted in a new agreement between the Settlement and the Chinese municipality of Shanghai, which in effect restored the court to its original status as provided by treaty. Procedural regulations were overhauled in favor of the Chinese, and once again the court formally became a Chinese-operated body, before which only treaty-power nationals could have a representative of their government join the Chinese magistrate on the bench. It was carefully labeled a "Provisional Court," which could in no way be construed as compromising China's campaign to abolish all extraterritoriality.[25] In fact the reformed court remained in force until all of these privileges were abrogated under the Nationalists. At the time Ting himself gave the new arrangement its modest due: "Although the full political aspirations of the Chinese people have not been satisfied in this agreement, as much was accomplished as was possible in view of the circumstances of civil war and the lack of a central government in the country." [26]

The problem of extraconcessional roads was very similar, arising out of the anomalous status of the International Settlement. As Shanghai had grown, Settlement authorities had succeeded in building roads outside of concession boundaries; once they were built, the Municipal Council claimed the right to collect property taxes from these areas and to police them. Jurisdictional disputes between the Chinese and foreign authorities, and even clashes between their respective police patrols, were commonplace, and current roadbuilding plans were at a stalemate. Although Ting entered into some preliminary discussions of these differences, he left office before negotiations could begin seriously. His strategy, however, was the same as with the court — to fight for a restoration of Chinese authority through a strict, rather than a lenient, interpretation of existing treaties.

In addition to negotiations with the Settlement, Ting's program for Shanghai also included ambitious plans for the modernization of the city's administration. He urged citizens to turn their energies to municipal improvement as the surest way to work for the eventual restoration of total Chinese sovereignty. Ting himself wrote the speeches which Sun Ch'uan-fang delivered in inaugurating a new

"Directorate of the Port of Shanghai and Woosung" under Ting's direction. One of them contained the following words:

Passing from the concessions to the Chinese city is like crossing the margin between yin and yang; the concessions are the world of yang and the Chinese city is the world of yin. Roads, buildings, and public health in the Chinese city in no respect compare with those of the concessions. This is the greatest of our national humiliations, far more shameful than the loss of sovereignty . . . One of the objects in establishing the Directorate of the Port of Shanghai and Woosung is to try and accomplish some real results to show our foreign friends and to start our general preparations for the future return of the concessions.[27]

The cure for foreign encroachment is not political resistance but administrative reform. The Greater Shanghai plan, authored by Ting, envisaged the consolidation of the various local districts of the Chinese city into a single administration. The whole would surround the existing international areas, hitherto the commercial heart of the city, and make possible the growth of a truly competitive Chinese metropolis supported by a new Chinese port. Similar projects had been discussed before, but they had always been obstructed by warlords and special local interests. Sun Ch'uan-fang's broad regional authority made it genuinely feasible for the first time since 1911 and gave Ting the freedom to declare the amalgamation of Nantao, Chapei, the Old City of Shanhai, Pootung, and the surrounding territories.

In the succeeding months Ting's task was to fashion this new system into a functioning unit, especially to create a modern police, health, and financial administration out of the patchwork of local authorities the directorate inherited, and to prop up the city's ramshackle tax base. He instituted proficiency examinations for employees of city offices and regular meetings of city department heads, and he recruited technical experts educated in Europe and America to take charge of new public health and public works departments. His reform plans extended to the development of a system for licensing medical practitioners, the creation of a professional police training academy, the enforcement of educational standards in city schools, and the establishment of a zoning authority. Outside of a vehicle registration tax, new sources of revenue were not to be found, but Ting did attempt to make up for this by instituting new property assessments and more efficient methods of collecting existing taxes. Even such a modest tax program was not easy to impose, but by October 1926, Ting was claiming that collections had risen by 20 percent. His plan was to use the new funds first of all for roadbuilding, in order to demonstrate Chinese efficiency to foreigners concerned over the problem of extraconcessional roads and to give the enlarged city an integrated communications network.[28]

Ting's brand of pragmatism was evident in all these initiatives. The belief that political or social choices are made on the basis of comprehensive factual data produced a campaign for vital statistics and a suspension of judgment about Shanghai's labor problem until Ting could master systematic information on the working classes. The structure of government which seemed practicable to him was authoritarian, and Ting worked under the powers delegated him by Sun Ch'uan-fang as the head of a directorate of city departments. His administration devoted itself to what would be the simple material necessities of modernization under any ideological banner: taxes, roads, health, harbor development, police. Politics was an unwelcome diversion of energy: "In a situation such as the

Shanghai Municipality finds itself the avoidance of political conflicts makes for increased efficiency, for the time and energy and expenditure which go into political controversy may be more advantageously devoted to the performance of administrative tasks." [29]

It followed that Ting's administration was not sympathetic toward the direct action policies of the effervescent street and student unions. Demonstrations on the 1926 anniversary of the May Thirtieth incident were discouraged, though not suppressed.[30] Renewed strikes in Japanese mills in June 1926 resulted in an order by the Chinese police to shut down the Shanghai General Labor Union and in an increased campaign of suppression against all "unregistered" unions. Ting did meet with mill owners to urge them to end a lockout, and he promised the Shanghai Student Union that he would try to curb the rice inflation which was exacerbating workers' grievances. But in fact he had no alternative but to rely upon the traditional half-measures of badly enforced price ceilings combined with charity sales of discounted grain by the Rice Merchants' Guild — practices which neither provided effective relief nor erased the prevailing opinion that speculation by these same merchants lay at the root of the problem.[31] Orders against strikes and demonstrations came from the provincial authorities or directly from Sun himself, but the Shanghai Police Department, which administered these orders and had wide discretionary powers, was directly under Ting's office. Moreover, Ting was clearly committed both by his administrative bias and by the temper of his negotiations with the British to a reduction of political and anti-imperialist agitation.[32]

In fact political passions rapidly engulfed his administration, in their most dramatic form as an armed struggle to seize power. In June 1926, within a month of Ting's arrival in office, Chiang Kai-shek led the Kuomintang armies north from Canton in the first stage of the Northern Expedition, which ultimately was to succeed in reestablishing a united administration over both north and south. As the summer advanced it became clear that Sun Ch'uan-fang, holding the strategic central Yangtze delta and in control of one of China's largest arsenals, stood between the rival armies of the Kuomintang and the Peiyang militarists. Officially his policy was one of peace, and in late July he called for a truce between Chiang Kai-shek and Wu P'ei-fu and for a solution to the problem of unity through a new coalition cabinet. But it was obvious he could not long delay a military commitment to one side or the other, and there was general speculation concerning his intentions. At first many thought that Sun was likely to reach a settlement with the Nationalists in Canton. A Kuomintang catchword was reported to be "Smash Chang [Tso-lin], defeat Wu [P'ei-fu], and accommodate with Sun." [33] In Hunan, Sun did aid the Nationalist cause by remaining aloof from Wu P'ei-fu, his old chieftain. The neutrality of the sixty thousand troops in Sun's command probably made the difference between victory and defeat for Wu at the pivotal battle of Ting-sze-chiao, which laid Hunan and Hupeh open to the Nationalist army.

Nevertheless Sun still continued to insist to Chiang on a policy of mutual noninterference, perhaps hoping that his rivals might destroy each other, or else that his friendship would be sought at a high price. The Communist-Kuomintang schism undermined his gamble, for once the revolutionary left had slipped out of Chiang's control in the Hunan-Hupeh vanguard, Chiang needed to win the coast and its arsenals in order to outflank the Communist-led wing in the interior. In early September, Kuomintang troops moved on the offensive in Kiangsi and Sun Ch'uan-fang found himself at war.[34]

To this crisis Ting responded with the administrator's perennial concern for order as a precondition of his own functioning. When the Kuomintang armies pushed north, he acted loyally to help Sun preserve an increasingly untenable independence. This involved him in military action against defectors from Sun's standard in the Kiangsu and Chekiang region and in repression of pro-Kuomintang groups inside Shanghai. The believer in administrative centralization found himself defending a hopeless policy of regional autonomy.

Beginning in early September the military crisis led to an increasingly tense situation within Shanghai. The directorate's relations with the radical workers and student unions, never warm, grew more strained as the authorities began to refuse permission for public meetings and processions. Ting and the police department were put in charge of planning city security; orders under Ting's name went out for special patrols to watch the arsenals and strategic roads and waterways, for a partial curfew, and for the imposition of city authority over police and militia in the outlying suburbs.[35] Inside the city civic groups began to clamor for peace,[36] radicals started to protest the suppression of the popular movement, and ordinary citizens grew increasingly resentful of the restrictions of martial law and anxious about their safety.

The climax came in mid-October when a surprise military attack against Shanghai was launched by defectors from Sun's camp in Chekiang, allied with revolutionary groups inside the city itself. The Chekiang governor, Hsia Ch'ao, at the head of an ill-armed force of twenty thousand men, was reported to be advancing on the city on October 16. On Ting's orders the Shanghai-Hangchow railroad track was torn up outside of Lung-hua in order to obstruct the invaders. As the authority in charge of the city's security forces, Ting collaborated closely with General Li Pao-chang, who was dispatched hastily by Sun to the city with three thousand men; together they organized the counterattack which dispersed Hsia Ch'ao's forces. Inside Shanghai a poorly organized street rising was put down by local police on October 24.[37] After these events the city remained under the special supervision of General Li as the commandant in charge of enforcing martial law.

However, this check to the Nationalists in Shanghai could not prevent the continued erosion of Sun Ch'uan-fang's power. Although Chekiang was restored to a state of uneasy loyalty for a few more weeks, other disasters accumulated for Sun. A series of defeats in the neighborhood of Nanchang and Kiukiang drove him out of Kiangsi, while Ch'en I, now the most powerful military figure in Chekiang, showed signs of encouraging a Chekiang "independence" movement which covered gestures toward a rapprochement with the Kuomintang. The possibility now loomed that Sun would attempt to retrieve his position with the help of new allies — from the north. Rumors spread that Sun planned a last-minute entente with the Fengtien armies and Chang Tso-lin, and they were fed by an ominous visit which Sun paid to Tientsin the week of November 21.[38] In fact an alliance with Chang was secretly being made.

Inside Shanghai the situation deteriorated rapidly. Fear of the "bandit" armies of the north caused disaffection with Sun to spread to the city's bourgeoisie, which had supported his policy as long as it meant peace. Once again an issue had created mass solidarity between the revolutionary street and labor unions and large sections of the merchants and gentry. On November 28, police were powerless to prevent a "Peoples' Assembly to Oppose the Movement of Fengtien Troops South" from holding a noisy demonstration. Blocking downtown traffic, fifty thousand people gathered to hear protesters denounce Fengtien militarism and call for united action for peace. Ting's old associate Ts'ai Yüan-p'ei took the chairmanship of a "Federation of the Three Provinces," which revived the campaign for local and regional self-government — expression of a pathetic hope that Shanghai could be immunized against civil war. A mass meeting on December 12 at the Shanghai Chamber of Commerce brought together these organizations plus representatives of the Kuomintang and the unions to denounce the suppression of civil liberties under Sun and to demand "citizens' autonomy." Talk spread of a tax boycott, of a general strike.[39]

Under such circumstances the society to "save Kiangsu" also knew that it preferred a future with the Kuomintang to one at the mercy of Chang Tso-lin. Opposition to the southward expansion of Fengtien power had, after all, been what had originally brought the society into being. Liu Hou-sheng (a member of the group who had not taken office under Sun) recollected later that the public announcement of Sun's alliance with Chang Tso-lin was directly responsible for Ting's resignation from Sun's government. On the advice of Ch'en T'ao-i, Liu and Ting went on December 11 for a final confrontation with the warlord. Liu recalled that Ting remained silent while he himself expostulated with the general for betraying the gentry of Kiangsu, who had looked to him for protection from "bandits."[40] But Fu Ssu-nien, with whom Ting discussed the episode later, wrote a slightly different account:

Sun, who was lying on an opium smoking couch, said, "Originally I felt as you do [about the Fengtien clique], but please read this telegram." The telegram had been sent by Sun's commander in charge of the front lines at Wuhsieh. In essence it said: we hear rumors that the Federated Command [Sun] and the Red Army [KMT] are planning to cooperate. At this news the troops grew dispirited. The Red Army are southerners, we are all northerners. It will be an evil day when northerners are subjugated by southerners, for they surely will take advantage of us. If there is no other way, then northerners must unite, etc., etc., . . . When Sun Ch'uan-fang had given them the telegram to read, he said, "I must ally with the Changs. Otherwise I am lost." Ting said, "If you ally with the two Changs, politically you will not have a leg to stand on." Sun said, "I can't take care of so many things." [41]

In the course of their drive back to Shanghai after the conclusion of this interview, a fortuitous automobile accident occurred which placed Ting in the hospital with a broken nose. He did not return to his office. Chiang Pai-li and Ch'en T'ao-i also resigned their posts, and Shanghai was soon in the hands of the Kuomintang.[42]

Ting, however, still clung to some remnants of faith in Sun's integrity, as may be seen by one final incident. The Kuomintang's military successes inexorably had the effect of pushing Chang Tso-lin and his allies into closer relations with the Japanese. Sometime in 1927, Ting, who was then living quietly in Peking, made use of the Chinese foreign office secret code — apparently through his connection with Lo Wen-kan — to send Sun Ch'uan-fang a secret telegram. He urged Sun to think now of his country and to break with the Changs. The telegram was intercepted by the Fengtien authorities and both Ting and Lo had to flee. Ting retreated to Dairen and lived in seclusion.[43] He had seen the man upon whom his nationalist ambitions once relied reduced to a foreign entanglement that seriously compromised Chinese sovereignty. It was his final disillusionment with the "patriotism" of militarists.

Under Sun Ch'uan-fang, Ting's experiences had been typical of those of the many intelligentsia who wished to reach an accommodation with the Peiyang military custodians of the republic. To his friend Fu Ssu-nien, Ting later justified his role in Shanghai on loosely pragmatic grounds: "to reform the Chinese government, . . . one cannot wait for times to be ripe. When an opportunity arises, one cannot miss it." [44] In his work there he had consciously carried out pragmatic precepts: exploit existing conditions, avoid ideology, and focus upon practical goals. Existing conditions in Kiangsu had demanded intervention to save the region from war and to correct foreign abuses; ideology led to fruitless demonstrations against imperialism in place of successful negotiation; and the most practical of goals were those which served the city's physical needs.

Behind this rationale was a view of government which was far from "practical" — a view which pragmatism did not intellectually expose, but which subverted the ultimate effectiveness of the pragmatists' official careers. It has been suggested that the reason why pragmatism in China was theoretically vulnerable to such a paradox was that the intellectuals were persuaded that politics is a purely empirical science, in which programs are constructed out of the data of experience, just as scientists presumably construct theories out of experimental data, and they were thus misled into accepting uncritically as "data" many social and political patterns of traditional China. Ting continued to believe that government is a matter of enlightened, authoritarian officials operating through a bureaucracy which naturally controls policy far more directly than does the formal seat of sovereignty. This led to his attempt to bureaucratize military leaders and, as an official under a warlord, to try and make administration return to its controlling role in the political process.

But in the last analysis Ting's behavior may not have reflected inherited ideas of the structure of government because his pragmatism dictated it. His pragmatism merely allowed it, and perhaps it was chosen for this reason. As a man he, like many others, was a descendant of the gentry in a dislocated society no longer able to make a place for gentry functions. In a stable culture the environment provides a constructive channel for those psychic energies which its own patterns of family life and its cultural myths engender in the individuals born there. In China "to be an official" had been not a career but a way of life, around which men had built an entire social and ethical identity, which affected not just their social prestige but the deepest layers of their self-esteem. Ting suffered the tragedy of a class brought up to a function it could no longer perform. Through service under a warlord he tried to find a substitute for the pattern of fulfillment to which he had been reared.

There remains the question of Ting's relations with the Kuomintang and with the Communist party, both of which inescapably confronted him for the first time during his tenure of office in Shanghai. In 1926 Ting displayed no attachment to the Kuomintang stronger than that which made most Kiangsu gentry prefer it to the Fengtien militarists. Later the record of his misplaced allegiance, which was highly embarrassing to his friends, caused some among them who were more closely associated with the Nationalist government to claim that Ting had always been "sympathetic" toward the Kuomintang.[45] But Fu Ssu-nien reported that Ting said he had never been consulted by Sun concerning the latter's relations with Canton.[46] The record indicates that Ting was not seriously attached to the Kuomintang's revolutionary purposes, since he was willing to help fight a war against the Northern Expedition and to acquiesce in the suppression of local Kuomintang militants.

For the Communists he must have had even less sympathy, since the tone of their propaganda during his administration was extravagantly hostile toward him. It can be inferred that Sun and Ting's policy of peace and administrative progress had considerable appeal, because the party's Shanghai magazine, *Hsiang tao*, took the trouble to warn against the disillusionment awaiting those who trusted to "false reform." For the Greater Shanghai project the Communists had nothing but scorn. Readers were told that it was really a device to make the people submissive to the imperialists and to concentrate tax revenues in warlord pockets, and they were encouraged in tax evasion and in resistance to the centralization decrees. For Ting personally the Communists reserved their most withering epithets of "running dog" and "loyal slave." They judged him an example of that corruption of spirit bred by intimacy with the West, which made Chinese acquiesce in their own humiliation and even praise the powers imposing it.[47]

Ting, after all, remained culturally pro-Western and politically antirevolutionary; and during these years there was little in the public stance of the allied Kuomintang and Communist party to appeal to a man of such convictions. The Kuomintang's banners promised the millennium with the end of the unequal treaties and denounced imperialism in an anti-Western and even anti-industrial tone hardly less vehement than that of the defenders of the "national essence." These slogans seemed almost as extreme as the frantic denunciations which

Communist leaflets heaped upon all attempts at political reform not first purified in revolutionary furnaces. As long as the Kuomintang and Communist movements remained above all revolutionary, Ting preserved an administrator's preference for order and an administrator's belief in its superior efficacy.

Chapter VIII

The Independent Critic

The year 1927 was in many ways a low point in Ting's life. After resigning his post in Shanghai he had retreated to Peking for a few months, but the enmity of Chang Tso-lin soon forced him to move again, this time to semi-exile in the treaty port of Dairen. Although self-doubt and self-recrimination were not characteristic of him, he had to live with political disappointment and the painful practical consequences of political disgrace — unemployment and isolation from the scientific world of the capital. In addition he was seriously short of money.[1] A solitary friend was able to visit him during a steamship stopover in Dairen. He reported that Ting was not embarrassed by his reversals of fortune, but that his style of living was indeed cramped. His wife was sick, and they were living with his future sister-in-law, whom he had taught to assist his scientific projects by drawing maps.[2] A good deal of his time was devoted to completing his edition of Hsü Hsia-k'o's *Travels,* which was published in 1928.

Gradually, however, Ting found new opportunities for scientific work. Sun Yat-sen University in Canton had considered offering him an academic post, but for political reasons decided against it. Others in the south were not so sensitive to his disgrace with the Kuomintang, and early in 1928 Ting received an invitation to go to Kwangsi and to lecture at the provincial university's summer school. Once he had arrived, local authorities apparently persuaded him to do some mineral prospecting also. In this way he was able to spend almost ten months of the year 1928, from March to December, on a field trip through his beloved southwest, during which he accumulated data for some of his most substantial research papers.[3]

Renewed absorption in pure research restored some direction to Ting's life, but at the same time he was left depressed that his own scientific investigations had lacked solidity in the past and that his present prospects remained clouded. A letter he wrote to Hu Shih from Kwangsi was filled with gloomy reflections:

I am a completely "impulsive" [English used] man. When I feel suddenly full of enthusiasm, I am capable of not caring about anything; when the enthusiasm fades, I feel dull and listless. This is why it is not easy for me to accomplish great things either in scholarship or in enterprise. So right now I cannot be certain of my future; it depends upon whether I find myself capable of enthusiasm after I return north. If I can live peacefully and quietly in Peking, and, moreover, if the Geological Survey can continue to exist as an organization and I can use it, perhaps I will be able to concentrate upon research for a few years. Otherwise there is no way to manage it. You know my research can't be done unless I have proper equipment and properly trained men to help me. If I leave the library and laboratory, and if there are no Grabau and his students to help me examine fossils, no cartographers to draw maps for me, there is absolutely no way for me to begin. The Geological Research Institute of Shanghai and the Geological Survey of Kwangtung and Kwangsi are too immature, and of no use.[4]

Happily Ting's fears remained unfulfilled, for the months in Kwangsi proved to be only the beginning of several years of increasingly satisfying scientific work. By June 1928 the Nationalist armies had finally driven the Peiyang militarists out of their last strongholds and had established a new, united government under the Kuomintang party. The period of widespread civil war seemed at an end, and Ting was free to move back to Peking. Although both Peking University and the Geological Survey had suffered badly from neglect during the recent conflicts, he found the organizations intact and both old colleagues and new students at work.

Almost immediately Ting began to plan and organize with his former zeal. During his travels in Kwangsi he had formed the idea that it might be possible to construct a railway from Szechuan to the sea along a southern route, which would be shorter and cheaper to build than the long-planned route parallel to the Yangtze River and would have the advantage of facilitating the development of the backward provinces of Kweichow and Kwangsi. The new Ministry of Railways in Nanking was persuaded to sponsor a reconnaissance survey,[5] and with this aid Ting took the initiative in organizing a large-scale southwestern field trip. By dividing into several parties, the scientists attempted to explore large sections of Kweichow, Yunnan, western Szechuan, and Kwangsi. Ting's ambitious plan was to amass enough stratigraphical and paleontological data for a comprehensive analysis of the geological structure of the southwest and, at the same time, to map and classify the aboriginal populations of the region. The expedition stayed in the field for over a year, from the late spring of 1929 until the summer of 1930. It was marred by a tragic accident. Chao Ya-tseng, leader of one of the parties and a particularly promising young geologist, was killed by bandits while traveling through northwestern Yunnan, rendering abortive a segment of the research and plunging the remainder of the group into a somber mood. Ting, who was deeply affected, later undertook responsibility for the education of Chao's orphaned son.[6]

In the summer of 1931, Ting became a research professor of geology at Peking University. His position was one of a group of professorships established as an early consequence of the Kuomintang-sponsored reorganization of Peking University in 1930, in an attempt to restore the institution to its former brilliance. Ting soon became one of a small inner circle of academicians who wielded special influence in university affairs. His close friends Hu Shih and Fu Ssu-nien had been instrumental in drawing up the plans for a reform of the university and in persuading Chiang Meng-lin to accept the post of chancellor· "With the aid of many friends — particularly Dr. Hu Shih, Mr. V. K. Ting, and Mr. Fu Shih-nian [sic]," Chiang said later, "the university sailed smoothly with only an occasional trimming of her courses." [7]

Like many scholars who customarily have been burdened with heavy research and administrative duties, Ting found teaching an exhilarating change. For three years he lived close to the classroom, taking personal charge of field trips, laboratory work, and introductory lecture courses. In the recollection of one student, he was even something of a classroom performer:

It is most difficult for me to forget Mr. Ting's manner of lecturing to a class: the cigar grasped in his left hand, the piece of chalk in his right, his deep gaze, composed bearing and extremely animated speech. He often used humorous expressions to stimulate students' interest in research and to create a lively and relaxed scholarly atmosphere . . . His flow of gemlike talk was truly lucid and memorable, and easily made the students chuckle more than a little. At each such satisfying juncture Mr. Ting would drop his chalk, permit himself two fierce puffs on his cigar, and pull at his mustache with both hands. In this kind of free lecturing atmosphere questions arose together with laughter, tobacco ash flew along with chalk dust, and what was originally a rather dry subject changed into a lesson everyone loved.[8]

These were productive years in research as well. Between 1929 and 1934, Ting published five substantial technical papers, a considerable output for a man whose lifetime publications in pure science amounted to under two dozen titles. In addition he was largely responsible for a valuable contribution to general scholarship in China, the Shen Pao Atlas, which he, Wong Wen-hao, and Tseng Shih-ying published together in 1933.[9]

Ting was unlikely, however, to confine himself to pure scholarship indefinitely. Much as he personally enjoyed academic life — and friends like Hu Shih later insisted that these years of teaching were the happiest of his life — he had not abandoned his fundamental commitment to the political and economic modernization of his country and his belief that the nation's intellectuals had an important role to play in this development. With the relative stabilization of the Kuomintang government, the political feuds of the warlord period faded, and Ting's past misalliances also receded from public mind. Moreover, the lull which followed the formation of a Kuomintang administration in Nanking was far too brief, and in the fall of 1931, soon after Ting took up his professorship, the Japanese seizure of power in Manchuria stunned a public which had just begun to hope that the worst of the internal crisis had passed. From that time onward, Japanese pressure upon China's northern borders never ceased, and slowly the country started down the grim road to war. Once again the sense of a nationwide emergency made Ting and his intellectual associates feel compelled to come to direct grips with political issues. As they had done in 1922, in 1932 they also organized a journal of current affairs. This time they called it the *Independent Critic* (*Tu-li p'ing-*

lun). Although the *Independent Critic* resembled *Endeavor* in superficial respects, its fundamental editorial posture indicated a new wariness and sense of limitations on the part of its principal contributors.

Like *Endeavor*, the *Independent Critic* was born out of a small discussion group of friends in the academic world of Peking. Hu Shih recalled that the magazine's collaborators were drawn from the faculties of Peking and Tsinghua universities, whose members felt the need to make a patriotic contribution to the struggle against Japan. The initial suggestion to start a magazine came from Tsiang T'ing-fu, then a professor of history at Tsinghua. There were only a few participants at first: Tsiang, Ting, Hu Shih, Jen Shu-jung, Fu Ssu-nien, Chu Yao-sheng, and Chou Mei-sheng. Among others eventually drawn into the magazine's circle were Ting's fellow geologists and friends Wong Wen-hao and Li Ssu-kuang. Hu and Ting were the only collaborators with previous experience on *Endeavor*, and they were also the most skeptical about the advisability of a new journalistic venture. Hu Shih warned of the difficulties to be anticipated from possible government censorship and of how unlikely it was that a magazine would produce much in the way of concrete results. However, he and Ting allowed themselves to be persuaded, and beginning in May 1932 they once more undertook the responsibility of being the principal editorial organizers and leading contributors to a journal of independent intellectual opinion.[10]

Endeavor had boldly offered a political platform and, with its doctrine of "good government," had attempted to provide a rallying cry for informed opinion; the collaborators on the new magazine were more acutely conscious of the flimsiness of their intellectual solidarity and the *Independent Critic* did not presume to offer a consensus. "We do not expect to be unanimous in what we advocate," they wrote in an editorial in the first issue. "We only expect that each man will investigate China's present problems according to his own wisdom and in a spirit of fairmindedness . . . We do not hope to obtain united opinions, but only hope to arouse criticism and debate which shall be public spirited and based on facts." [11] Earlier their ideal had been participation and action; now they claimed to be content with the critic's stance of independence — from money, power, fashion, or prejudice — and they pledged themselves above all to the modern journalist's value of free speech.

Only occasionally did the writings of Hu Shih echo in a muted fashion the old rhetoric of struggle and work. Ting still believed in the special mission of the intelligentsia, but he no longer seriously looked to them as the nucleus of a militant political movement; he still believed in the educated elite, but his idea of their role had undergone a subtle change. True, in an early issue he asked that independent intellectuals "have firm beliefs, win the sympathy of youth, make concrete plans, and . . . unite small groups of those with likeminded beliefs to perfect a large political organization." [12] But this turned out to be more an echo than a new commitment. He still spoke of educated men, but he no longer mentioned "good men"; their duty now was to introduce modern technical and administrative systems, no longer to inspire the moral regeneration of the state. The contrast with *Endeavor* suggests a quite fundamental shift in the orientation of the academic intellectuals toward society. By 1932, Ting and his associates had reached another stage in the Chinese intellectual's evolution away from a Confucian notion of his own function. No longer did they easily assume the existence of an intimate association between knowledge and power which should theoretically place them, as educated men, at the center of the political process. In the *Independent Critic* they wrote more like members of a Western intelligentsia, as self-conscious outsiders and skeptical critics of society, to whom the alienation of the intellectual is an expected condition of modern life and impotence seems the natural companion of insight.

The *Independent Critic,* then, was simply what it said it was, a journal of opinion; ironically, though, it was more successful than the more ambitious *Endeavor* had ever been. A somewhat more favorable political environment helped. Instead of dealing with fragmented military governments intent upon civil war, the intellectuals were now responding to an administration formally united and at peace, which claimed to be the custodian of Sun Yat-sen's tradition of Republican reforms and which professed a genuine interest in modernization. By not claiming a major role for themselves, the intellectuals were more able to match their performance to their pretensions; given a temporarily stable structure of government, their "practical" criticisms were more likely to be relevant to existing possibilities. Though limited, the magazine's influence was genuine. Contributors often held government office, if not in high policy-making circles, at least in administrative agencies dealing with economic development, diplomacy, finance, or education. The magazine survived until the outbreak of the Sino-Japanese war in 1937, and it was widely enough read to be self-supporting.[13]

In his writings for the *Independent Critic,* Ting was overwhelmingly preoccupied with two related problems: Japanese aggression and Chinese industrialization, for the latter was now seen not simply as a developmental goal, but also as a prerequisite for all effective resistance to the national enemy. Together with most of his associates on the magazine, Ting opposed the idea of a Chinese declaration of war against Japan, and he called for the patriotic efforts of his countrymen to be concentrated instead upon an intensive program of economic and industrial development.[14] In the face of a rising mood of nationalistic militancy, especially among the student generation, Ting essentially supported the Kuomintang tactic of appeasement in northern China.

His views emerged quite clearly during the Jehol crisis of February 1933, when the Japanese moved down from central Manchuria to take the strategic mountainous border region overlooking the north China plain. At the height of the ensuing war scare, as Japanese troops marched to within a few miles of Peking, Ting bluntly told an audience of university students at the capital that to advocate all-out war was irresponsible. One by one he dismissed the expedients which patriotic emotion suggested to the young — anti-Japanese boycotts, military drill, propaganda drives, and above all, student enlistment in the army. War in 1933 would lead only to disastrous military defeat, he claimed, since China entirely lacked the industrial and economic capacity to win a modern military campaign. The development of such a capacity was nothing less than a question of the modernization of the entire nation, for which there were "no short cuts." Students represented the precious trained manpower capable of undertaking the hard work of construction, and the thought of such individuals quixotically risking death in battle struck him as the height of folly. Instead he urged students to undertake the sober work of training themselves as a skilled, modern-minded elite.[15] While he cautioned the youth on one hand, Ting at the same time criticized the Nationalist authorities, and especially the Manchurian army of Chang Hsüeh-liang, for their weak resistance within Jehol, where, he thought, the rough terrain should favor a resolute defending force.[16] After Jehol had been abandoned, Ting joined with Hu Shih, Chiang Meng-lin, and Mei Yüeh-han in a public letter calling for Chang Hsüeh-liang's resignation;[17] and Chiang Kai-shek himself, in a personal interview with top Peking academic leaders, offered Ting and others a few words of regret and explanation.[18]

Ting's position on Jehol did not change his basic opposition to a Chinese declaration of war, but simply reflected his feeling that there was some deterrent value in having the enemy pay as dearly as possible for its own initiatives in aggression. In thinking of the future he was willing to envision quite drastic losses of Chinese territory as the price that might have to be paid for the time needed to build a Chinese defensive capacity. In 1935 he offered an admittedly provocative comparison of China's predicament with that of Russia just before Lenin made the Peace of Brest-Litovsk. Dramatically abandoning the Ukraine to the Germans, Lenin had said that to save the Bolshevik Republic, he was prepared to retreat to the Ural-Kuznetsk region, or, if need be, even as far as Kamchatka. China's choices, stated Ting, were no happier: "North China is our Ukraine; Hunan, Kiangsi, and Szechuan are our Urals-Kuznetsk; Yunnan is our Kamchatka. I wish that our top leadership would study Lenin and determine what is the most important thing today. Everything else can be yielded. Our second-ranking leadership should study Trotsky: before the decision he bore responsibility and offered his views; and afterwards he harbored no grudges, claimed no merit, and cooperated as before. I wish that all of us would get ready to go to Kamchatka." [19] Ting knew these suggestions were controversial: although the *Independent Critic* on the whole favored appeasement, even the peaceable Hu Shih objected to the idea of a Chinese Brest-Litovsk. Actually, by the last months of his life, Ting no longer emphasized such an expedient. According to Tsiang T'ing-fu, Ting had privately come to doubt whether peace was possible for much longer, and many of his plans and actions reflected a quiet preparation for war.[20]

In considering the crisis of military defense Ting thought mostly in terms of the technical and organizational requirements for effective modern war; perhaps in self-conscious opposition to the rhetoric of militants, he did not emphasize the importance of morale or psychological commitment on the part of the Chinese people. By the 1930s Ting talked more like a materialist and less like a voluntarist than he had ever done before. For the long-term salvation of China he offered scientific modernization. In the environment of the 1930s this meant economic planning on the model suggested by socialist experiments in the West. Paper programs of "construction" (chien-she) and of a "controlled economy" (t'ung-chih ching-chi) were an important part of the Kuomintang's progressive facade, and the collaborators on the Independent Critic in general advocated "scientific planning" as the latest in that long line of Western techniques which had been proposed as shortcuts to national strength and prosperity.

In lecturing his audiences in the Independent Critic on scientific construction, Ting was above all haunted by the disappointments of the past and anxious to make clear that successful modernization required an attack on a broad spectrum of interlocking problems. In spite of the Japanese threat, narrow military priorities in planning struck him as futile, and he pointed out to Kuomintang officers how arms alone had not saved the Chinese forces against the Japanese in 1895. He emphasized the relevance to economic planning of political conditions like national unity, legal ones like the end to unequal treaties, administrative ones like the rationalization and centralization of local government, social ones like better education.[21] As for successful planning techniques, these depended on scientific methodology, whose nature was still imperfectly understood by the Chinese public. In language which echoed his basic positivist philosophy, Ting warned readers against a narrow view of the scientific enterprise:

In the learned world science includes everything. What is called "scientific" or "unscientific" is a question of method and not a question of subject matter. All phenomena and facts in the world are materials for science, and provided the method is correct, all may be acknowledged as science. What is called the scientific method is to take phenomena and facts and use the logical method to classify them systematically; to understand their mutual relations, look for their common principles, and predict their future consequences. Therefore we may say that this kind of knowledge is true, which is equivalent to saying it is scientific. We say that a project is systematic and rational, which is equivalent to saying it has been made scientific.[22]

But the scientific social planning of the academicians, with its new imitation of Western formulas for socialist construction, did not win favor. Rather, Chinese radicals had learned the Western Communist left's skepticism concerning reforms proposed in a reactionary political and social environment. To such people the academic leadership appeared socially compromised, and pro-Communist Peking students claimed that the science-minded intellectuals were merely advocating the development of a native capitalist system, in which sophisticated methods of production would be turned to the work of social exploitation. On the other hand, to young patriots the slogan "Save the nation with science" seemed narrowly specialized and ineffective, and its proponents were tainted as appeasers. "I feel that the words of the *Independent Critic* do not satisfy our craving," a typical young reader wrote about the magazine's articles on Manchuria.[23] Ting was concerned by the gulf which he saw growing between his own and the younger generation. Their attraction for Communism appeared entirely natural to him, and he fretted that the recommendations of older men like himself risked seeming even less palatable when they were accompanied, as they all too frequently were, by a failure of sympathy for youth.

Every one of the issues which he discussed in the *Independent Critic* — Japanese aggression, economic development, Kuomintang policy, the allegiance of youth — at some point involved Ting in a consideration of the question of Communism in China. By the early 1930s he had come to believe that the Soviet Communist regime represented an experiment in social organization of epoch-making importance and that Communism inside China was a major political movement and a formidable attraction for the uprooted younger generation. His writings show that with his usual methodical scholarship he undertook to found his own view upon the study of Marxist doctrine, Russian history, and Soviet economics. Furthermore, he dramatized his curiosity and concern by undertaking a two months' journey through the Soviet Union in the summer of 1933, about which he offered a detailed report afterward in the *Independent Critic*.

Even before he left on his trip to the Soviet Union, Ting published an essay on Communism in the *Independent Critic*.[24] With a typical intellectual tendency to judge a political movement on the basis of the abstract doctrine professedly at its basis, Ting turned first of all to the problems of theoretical Marxism. His views were close to those of academic critics of Marx among contemporary Western sympathizers with the left. He pointed out the well-known inadequacies of a theory of economic value based solely upon the value of labor; and he expressed the familiar distaste of scientific empiricists for Marx's doctrine of "dialectical materialism" as a scientific philosophy of history. "History," Ting stated, "has no logic," and he maintained that to believe the contrary was a misuse of the idea of science. However, it appeared to him that Marx, in spite of his intellectual errors, had shown a profound intuition about broad social trends in nineteenth- and twentieth-century Europe. The vision of "equality," born of the French Revolution, had taken powerful hold of men's imaginations in an era when developing capitalism produced economic conditions particularly hostile to this ideal: "Marx bitterly hated social inequality, and so he wanted to stir up class warfare. Because he wished to stir up class warfare, he created his labor theory of value to prove that workers had lost their property rights and that capitalists are thieves. His theory of value was said to be an economic principle; it would be better to call it a political slogan." [25] As a political doctrine calling for economically equalitarian societies, Marxism deserved its worldwide appeal, and in this respect its goals ought to be shared by all modern governments. For Ting the ideal of equality seemed entirely praiseworthy when presented in the socialist guise of economic sharing rather than the democratic one of the equivalent spiritual value of all individuals; moreover, he was convinced that economic equalitarianism was a practical necessity in the contemporary world, where the existence of privilege inevitably led to political discontent.[26]

When he turned from doctrines to men, Ting alternated

between two contrasting inclinations — admiration for the performance of the Russian Communists, whom he saw as political leaders in power dealing with the problems of a weak and underdeveloped nation, and skepticism concerning the pretensions of the Chinese Communists, whom he saw as revolutionary demagogues spinning a rationale for insurrection out of the speculations of a pseudoscientific theory. He feared the Chinese Communists were little more than men dedicated to violence and terrorism, who used Marx to justify themselves as the catalysts of social change by means of bloodshed and to sustain an apocalyptic faith in their own inevitable success. On the other hand, Bolshevik leaders like Lenin and Trotsky had proved themselves remarkable manipulators of political power; they had successfully disentangled a weak and foundering nation from a disastrous foreign war, fought widespread civil rebellions to achieve internal unity, and created an honest administration and a centralized planning apparatus capable of rallying Russia's resources for industrial development. By implication the Kuomintang's performance in the face of many similar problems had been a failure.

Such was Ting's perspective when he set out for the Soviet Union in the summer of 1933. He was probably one of the few Chinese of his generation who visited Russia as a relatively established and cosmopolitan observer whose perspective had been influenced by a European education and a familiarity with the workings of mature industrial systems. Ting reached Russia by way of the International Geological Congress in Washington, D. C., which he participated in as an official delegate in July 1933, and after several stopovers in Europe to visit old friends in England and Sweden and his brother in Germany.[27]

In this way Ting obtained his first real glimpse of the United States, which he had seen before only in transit on his journey to Europe in 1919. In 1933 his stay included visits to Chicago, Washington, and New York, and his impressions form an interesting contrast to those he wrote about during his visit to the Soviet Union soon after. Ting's principal reaction to America was one of dazzled amazement at the wealth and inventiveness of its material culture. The absence of detectable signs of poverty in spite of the prevailing economic depression, the speed and comfort of its transport, the splendor of its scientific museums, and the lavishness of its scientific research establishments all contributed to an impression which he expressed most eloquently at the spectacle of the New York skyline:

As the ship was leaving the wharf, I looked at the tall buildings beside the Hudson River and an indescribable feeling came over me. On the one hand it was fear, . . . causing me to understand the power of material culture and at the same time to be aware of the limits of an individual man. On the other hand it was appreciation: there were not one or two of these buildings, but dozens and hundreds, high and low, ranging along the shoreline. The old were perpendicular like rectangular boxes; the new were shaped like pagodas — they rose straight up to a suitable height, then story by story tapered to a point . . . These are products of the twentieth century, works of art of the age of steel and cement, and how can wooden or stone buildings be compared with them? Many Americans point to New York's "skyscrapers" and even call them ugly, but to prostrate oneself instead in homage before the churches and palaces of Europe's Middle Ages may indeed be called a foolish self-deprecation — to love the old and belittle the new! [28]

Ting felt a temperamental attraction for America as a young, spacious country lavish in its applications of modern technology. However, he did not attribute the American phenomenon to any qualities in American economic or political institutions. Where he commented at all, it was merely to note with the perplexity of one to whom etatism seemed the natural order of society that the United States government appeared remarkably indifferent to domestic education, science, and culture. To Ting, America was a marvelous freak of nature: a country which, like a beautiful woman, enjoyed rare opportunities through the sheer splendor of physical endowments. He admired much about it, but it did not occur to him to see in it a model for China to imitate.

It is perhaps regrettable that in writing the chronicle of his Russian journey Ting did not feel the obligation usually displayed by European visitors to that country to describe their experiences in a frankly political context. The impression of factuality and restraint which characterizes Ting's account may derive partly from the fact that the chronicle is incomplete: he wrote it in weekly installments for the *Independent Critic* and he had only covered the events of one-half of his two months' stay when other business and, unexpectedly, his death intervened. But it was in accordance with his political creed as well as his personal style in journalism for him to write as if the facts should speak for themselves with the objectivity of a naturalist's field notes. For the most part Ting's feelings about life in the Soviet Union must be inferred from the selection of facts which interested him.

Ting's host was the United Geological and Prospecting Service (Soiusogelorazwed), the Russian counterpart to his own Geological Survey and an enormous organization with regional branches all over the country. Under the First Five-Year Plan, then in midstream, the regime was attempting to bring academic science into close contact with

industrial and development problems. Ting found geological work in the midst of a period of rapid growth. Over
the preceding several years a 90-percent expansion program
had been implemented, making the United Geological
and Prospecting Service into an impressive organization:
it included a research institute in Leningrad, sixteen provincial offices, and seven schools to train various kinds
of geological specialists; and it employed three thousand
full-time geologists, in addition to two thousand field
workers and five hundred cartographers. Its annual budget
was one hundred and twenty million rubles. Ting estimated that it was the largest geological research organization in the world, and he emphasized the coincidence that
statistically speaking it could be described as exactly a
hundred times the size of China's Geological Survey.[29]

Wherever he stopped he saw the same spectacle of an
intense drive for industrialization placed side by side with
signs of poverty which illustrated the background of deprivation against which so much was being accomplished.
Ting's standards of affluence were set by Europe, and he
did not fail to consider the conditions he met with on
much of his trip as shabby and primitive. Outside of the
cities the physical facilities of a large part of this supposedly Western nation in fact reminded him of China:
highways rapidly degenerated into dirt or cobblestone
roads on which the rare government automobiles regularly broke down; buildings were dilapidated and flimsily
constructed; traveling accommodations were dirty and frequently bedbug ridden. A dining room at a cooperative
farm where he ate lunch, with its dirt floor, wooden
benches, and intimate view of the stable outside, reminded
him of the filth of roadside teahouses in Yunnan; a railroad journey to a Moscow suburb in a bare third-class
carriage jammed with marketing peasants made him think

of the Peking-Suiyuan railroad during one of China's civil war panics.

But conditions which by comparison with Europe were poor indeed, by comparison with China served to suggest what planning might accomplish in spite of environmental handicaps. The dislocations that were the result of overly rapid growth pointed to the same lesson. Ting noticed that the quality of Russian geological work frequently suffered because positions of great responsibility in the field had to be given to relatively inexperienced men, but he also noted that the organization knew this and was trying to correct it. At Tula, where the iron and coal fields had just recently been slated for expansion, new drills had only begun to supplement the hand-cutting of coal, while the product continued to be transported by donkeys and pulley carts. The iron mines had received up-to-date smelting furnaces only four months before his visit. Everything in fact suggested rapid evolution from conditions not unlike those of the Pei-p'iao mines. The Georgian University at Tiflis possessed a substantial library of foreign books, yet it had been unable to train its students in the foreign languages required to read them. But at the same time Ting also expressed amazement that this twenty-year-old university was providing an education for seven thousand students from a stratum of society which had always been condemned to ignorance before. In Russia, Ting saw being carried out the kind of projects for the modernization of a society that had populated his dreams for China all his adult life. Moreover, he saw this being done in a nation where the material conditions of the masses bore some resemblance to those of the Chinese themselves.

Furthermore, in the field which he personally understood best, scientific research, his experiences in no way confirmed anti-Communist claims that despotic and politically oriented supervision of intellectual life was stifling fruitful scientific work in Russia. He found Russian geology in a state of vigorous expansion and, even more important, Ting communicated with his colleagues in Russia in an atmosphere of cosmopolitan cordiality. He found the members of Leningrad's Geological Research Institute well acquainted with his own work and that of his Chinese colleagues. They talked freely in a mixture of French, English, and German about commonly shared interests in the geology of north China and of the Asian continental heartland. In their company Ting felt an ease which, he said, "made me feel that science is international." [30]

Considering the problem of freedom in Russia, Ting's reservations amounted to little more than a lingering feeling that he personally would always be more comfortable in the bourgeois intellectual atmosphere of Europe or even America. He expressed this feeling in a little game he played to amuse himself on the train coming home from Russia: "I asked myself on the train, 'If I were free to choose, would I rather be an English or American workman or a Soviet intellectual?' Without hesitation I replied, 'An English or American workman.' Then I asked, 'Would I rather be a Paris White Russian or a Soviet geological technician?' Also without hesitation I replied, 'A Soviet geological technician.' " [31]

As his game on the train hinted, Ting imagined that the Soviet experiment fulfilled for Russia what were two of China's gravest needs: it provided a central government which was identified with nationalist aspirations, and it offered a method for technical progress. His trip to Russia confirmed his estimate of the Soviets as pioneers in the scientific engineering of an entire society, of a national government entirely committed to scientific progress and ready to make creative use of the technically skilled individuals who were most qualified to bring it to fruition. Writing from the perspective of Taiwan in the 1950s, Hu Shih lamented his late friend's "prejudices" on the subject of Soviet Communism, which Hu thought had been largely influenced by the sight of government-sponsored geological progress in Russia.[32] After his return home, Ting felt the attractions of Communism enough to prompt himself publicly to ask the question "Why am I not a Communist?" and to try and give it a formal answer.[33]

Basically he remained a non-Communist, not so much because of his intellectual reservations concerning Marxist theory — though these did influence him — but because the national strength and technical progress which he saw Communism bringing to Soviet Russia seemed to play too small a part in the prospects and programs of the Communist party inside China. An armed political movement on the fringes of real power, it remained revolutionary, fanatical, and divisive. In the Soviet Union rationalized planning and construction in accordance with scientific principles seemed a reality, whereas in China science served as little more than the propagandistic rationale behind a theory of revolutionary war. Moreover, the Chinese Communist leaders in no way corresponded to Ting's image of a competent, scientific-minded elite. Significantly, he accused them of the most serious of all possible defects in his private political lexicon — of being undereducated. From their writings they impressed him as provincial, superficial, and slavishly deferential to the slogans of Marx and his followers. He suspected that they possessed neither the necessary Western learning nor an intelligent common sense approach to modern organizational methods which might give their plans for construction a chance of success.

It also seemed to Ting that the Chinese Communists put ideological theories and revolutionary ambitions before China's national interest. He never saw them as champions of the United Front against Japan, and he assigned responsibility for the Kuomintang-Communist schism equally to both sides. Communist determination to persist in secret insurrectionary activity was, he maintained, no more in the national interest than the Nationalist drives of "bandit extermination" at the expense of China's northern defenses. He accused the Chinese Communist party of further dividing the nation at a time when it needed above all to be united, and of being reckless in policies which would alienate Western powers at a time when their presence in China was an important weapon in the struggle for national defense. With the pathos of reasonableness in such conflicts, Ting asked only the impossible: that the Communists change into a non-violent, open, opposition political party, and that the Kuomintang tolerate their existence as such.[34]

Invariably a man of order, Ting found it far easier to sympathize with a Communist bureaucrat than with a Communist revolutionary. Looking at the Communist movement in 1935, he underestimated the ideological dogmatism and revolutionary zeal which animated Bolsheviks like Lenin and Trotsky in the years before 1917, nor did he anticipate the organizational skill which the Yenan leadership would muster once it had secured power in Peking. In spite of his equalitarianism, Ting also betrayed some sense of social estrangement from the Chinese Communists — their rough-and-tumble backwoods talents seemed alien to an intelletcual who had confidence primarily in the education of the schools, and their immersion in the intimate daily life of the Chinese peasants had never been approximated by a scholar who thought that "Chinese conditions" were best understood through a study of Western science. But his deepest estrangement was from the mind of a revolutionary, who risks and even welcomes violence as a necessary prelude to justice and order in the future. Rather than plunge Chinese society into further anarchy, Ting felt that almost any pragmatic makeshift was preferable: "Why then am I not a member of the Communist party? First, I do not believe that revolution is the only way; and I especially do not believe in any 'logic of history' able to guarantee a revolution's success, or that in any kind of environment a revolution will necessarily take the same form. Second, I do not believe that outside of continuous long term struggle there is any shortcut to human progress. Therefore I sympathize with a portion (perhaps a great portion) of Communism, but I do not agree with the Communist style revolution." [35]

The spectacle of Communism in power in Russia, compared to that of Communism in the rural areas in China, undoubtedly influenced Ting in a political view which he proclaimed at the cost of some controversy in his last years — his explicit repudiation of democracy. After the stabilization of the Kuomintang government in Nanking, a revival of the campaign for the democratization of politics in China occurred, led mostly by Western-oriented intellectuals, who hoped that the Kuomintang would now try to live up to some of its democratic professions, and by Kuomintang dissidents who resented the concentration of power in the hands of the military circle around Chiang Kai-shek. On the other hand, there arose questions of how far democratic methods could succeed in accomplishing some of the necessary objectives of the regime, which included pacification of local areas and the subjection of remaining independent warlords, as well as the related consolidation of the armed forces under national command. In 1934 and 1935 the collaborators on the *Independent Critic*, together with contributors to other established moderate journals like the *Ta kung pao, Kuowen chou-pao*, and *Tung-fang tsa-chih*, wrote a large number of articles on the subject of democracy and dictatorship, which were held up as two alternative model forms of government for China.

On the surface this debate had a curiously academic air, since participants spent most of their time attempting to draw lessons in political science from the facts of Western history by offering contrasting estimates of the applicability of European and American political institutions to Chinese conditions. Underneath there could be discerned certain infinitely circumspect criticisms of the Kuomintang practice of "political tutelage" — of its authoritarianism by one group and of its inefficiency by the other. In fact the debate was precipitated by a veiled Kuomintang attempt to defend itself from the charge of despotism, in the form of official statements that "Chinese conditions" — that universal scapegoat — did not permit the establishment of a fascist government inside China.

Thus Kuomintang performance supplied one underlying theme of the debate. A second theme was compounded of Chinese responses to the disturbing rise of totalitarianism in contemporary Europe. Western democratic institutions, whose unsuitability for China had already been claimed by conservatives in the 1920s, now appeared as blighted with the most serious of stigmas in the eyes of an experimenting Asian nation — failure. Chinese democrats found it even more difficult to advocate for Asia a form of government unable to thrive on its own soil. When pressed, prodemocrats like Hu Shih, Wu Ching-ch'ao, Ch'en Chih-mai, and T'ao Meng-ho largely confined themselves to defending democracy as a theoretical ideal and to suggesting that the Kuomintang be encouraged to evolve gradually in a more libertarian direction.[36]

The most forthright advocates of dictatorship who were not merely acting as apologists for the Kuomintang were Ting and Tsiang T'ing-fu. Tsiang, himself a historian, argued that historically proven methods of national evolution required the preliminary consolidation of national territory by a monarch. Since China's emperors never completed this historical task, a modern dictatorship was now needed to establish a strong nation. Only later would it be appropriate to consider a transition to gentler forms of government designed to make the people happy.[37]

Ting also argued for dictatorship from historical necessity, though in a less abstract fashion and not without some equivocation. Although he in no way admired either Hitler or Mussolini, he believed that their successes, together with that of Soviet Russia, proved that democracy had little hope of surviving in nations which did not enjoy long historical conditioning in libertarian forms of government. Moreover, he was sensitive to the mood of disillusionment which the spectacle of fascism had created among the European democratic intelligentsia. By the 1930s his favorite British authors like Bertrand Russell, Harold Laski, and H. G. Wells were beginning to question whether the European democracies were capable of solving the severe twentieth-century problems of economic privilege, nationalist rivalry, and technological change.[38] Ting spoke of the apathy of democratic electorates, of their incompetence in making political judgments, and the ease with which they could be manipulated by press, power, and money. In these criticisms he was echoing the pessimism of the British Fabians in their old age, for whom the democratic machinery created in Europe's nineteenth-century innocence now seemed too flaccid to cope with economic inequality at home and fascism abroad. For Ting the conclusion was obvious: where mature societies enjoying universal education, developed communications, robust political parties, and a prosperous economy could not maintain stable democratic rule, China, beset by aggression from Japan, facing a domestic economic crisis, and lacking all historical conditioning for republicanism, dared not hope to try.

Ten years after he had first become involved in politics through *Endeavor* magazine, Ting felt more strongly than ever that the only solution to China's political crisis lay through the instrumentality of an enlightened dictator. But where before he had made an appeal to vaguely delineated military heroes and "good men," now he demanded a "new model" (*hsin shih*) dictatorship, which he described as follows:

1. The leadership of the dictatorship must consider the nation's interests identical with its own.

2. It must thoroughly understand how to modernize a nation.

3. It must make adequate use of the nation's trained manpower.

4. It must use the nation's present difficulties to rally behind its own banner the allegiance of all people capable of participating in government.[39]

This formula maintained the basic structure of government which had always most deeply appealed to Ting — ruler and an educated bureaucracy. But by the 1930s the content of this structure had slipped even further from its Confucian moorings, so that Ting insisted that the ruler be a "new model" one (Stalin? Roosevelt?), and that the elite conform to the mid-twentieth-century socialist ideal of scientific government planners. No longer did it seem plausible that the necessary requirements for rule might be fulfilled by a Wu P'ei-fu or a Sun Ch'uan-fang — the ineptitude of China's military leadership was being proven again by Chiang Kai-shek, symbol of what Ting called "the old style despotism in a new mask." [40] In spite of his desire to support the Kuomintang as long as it was the only available nucleus for a stable administration, Ting's writings in the *Independent Critic* expressed his skepticism concerning the Nationalists' pretensions to modernity and efficiency. In addition to its inadequate defense measures and imperfect regional control, it had shown itself unwilling to tolerate opposition, incapable of attracting the loyalty of youth, ignorant of modern planning and administration, narrowly militaristic, and venal. Ting's idea of an enlightened dictatorship was now drawn from broadly yet indistinctly perceived Western models: from the America of Roosevelt's New Deal or the Russia of Stalin's Five-Year Plan. In spite of their ideological conflict, Ting persisted in viewing these two governments as representative of a fundamentally common global political pattern. They were governments in which authoritarian leadership worked hand in hand with an administrative elite, a corpus of technically trained bureaucrats in whose hands necessarily lay the increasingly sophisticated operations of economics, industry, education, and finance, by which alone a modern state can function. Consistent with his Confucian heritage, Ting continued to imagine that the ruler and the educated bureaucracy are the structural pivots of any political system, but by the mid-1930s he had infused into these forms another final change of meaning: the elite had become a technocracy and the ruler the director of a brain trust.

His views of Communism, of the USSR, and of dictatorship all illustrate Ting's lack of central concern for political liberty as it is understood in the West. He moved from an elitist and instrumentalist view of the democratic process in the 1920s to the frank advocacy of enlightened dictatorship for China after 1933. Like other Chinese students of Western liberalism, popular participation in government seemed relevant to him when he considered the problem of how to rule effectively, and the conclusion that state goals would not be achieved through democratic methods cast doubts upon the system in principle. By contrast most Western liberals of the nineteenth century and afterward have taken the position of the individual as the theoretical starting point for deliberation, and methods of governing have been considered above all from the point of view of how they relate to this central value.

What was Ting's feeling about freedom itself? Tradition had bypassed the issue by linking knowledge and action, and Confucian statecraft assumed that no inherent conflict existed between an individual's highest values and state power. Rather, the optimistic presumption was that basic "principles" are embedded in man's nature, and that the task of both culture and government is simply to train these "principles" to manifest themselves in working social relationships. One might conjecture that as a man like Ting moved away from a Confucian conception of his function, the value of freedom would gradually emerge as more important to him. By the 1930s Ting's personal outlook approached that of a Western-style intellectual more closely than ever before, and in the twentieth-century West freedom may even be seen as the value of a class, the consequence of a certain structural position in society that modern intellectuals typically hold. Ting, however, was if anything even less "liberal" in 1935 than he had been fifteen years earlier. The reason lies, perhaps, in Ting's idea of science.

The only written comments he made about civil liberties were offered in answer to questions about freedom of scientific thought in the Soviet Union. Russia, he said, did not control scientific thought because in the last analysis, "scientific research cannot be controlled . . . science does not know 'authority' and cannot submit to 'authority's' supervision." [41] A government's commitment to scientific progress, he implied, necessarily involves its willingness to accept scientific reasoning and its conclusions. In scientific social practice as in the physicists' laboratory, the responses of men will be governed by the answers to which a correct methodology must lead. If not the knowledge of necessity, freedom for Ting came close to being the knowledge of scientific law.

In his focus on the administrative process and on technical modernization, and in his acceptance of authoritarianism, the Ting of the *Independent Critic* remained close to the man who had attempted to apply the formula of the Chinese pragmatists in his administration of Shanghai. But by the end of his life one of the fundamental axioms of that pragmatism had come to seem empty to him. In 1934, in a mood of broad historical rumination, Ting stated that after all what the Chinese most needed was an ideology. Even though he had no specific form of belief to propose, he saw clearly that practical solutions and the scientistic appeal to reason had failed to give the Chinese people the sense of purpose they needed to unite and solve the problems of national modernization.

Fundamentally, he observed, China had been unable to modernize because in the few brief years since she had been committed to technological and social revolution, her people and her leaders had never shared the most basic beliefs. The stability of nineteenth-century Europe and America could be traced to a set of broad common beliefs — in parliamentary politics, laissez faire economics, and the separation of church and state — beliefs which ensured domestic tranquillity and permitted the enjoyment of civil liberties. It was unfortunate, he reflected, that China had fallen under the influence of Europe not during that happy time, but later, in an age of political and intellectual conflict. Educated Chinese, products of Europe's contending schools, had cast China's revolution in the image of the schismatic West when, in fact, China could not survive without some common beliefs to rally and unite the entire people. The development of a common national faith, Ting concluded, was far more essential to national regeneration than the manipulation of specific policies or shifts of personalities in power.[42]

Although Ting ended this particular essay with an appeal to the Kuomintang and other parties to seek areas of common agreement, he no longer did so with the confidence that some pragmatic formula based on practical problems was likely to prove workable. Even more important, he did not suggest that scientific reasoning, however useful as a guide to administrative policy, was adequate to supply an ideal around which the Chinese could achieve a sense of common allegiance. In "Unity and a Common Faith" the voice of the scientist was stilled. What remained were the anxieties of a man who realized that successful systems of belief are not the product of scientific ratiocination, but of slow historical growth, and yet who knew that the Chinese nation did not have time to enjoy the luxury of gradual, organic evolution toward a new social order.

After 1931 Ting himself stayed on the perimeter of

political life. Occasionally he served as an economic consultant to the government. He was close to high officials in the Ministry of Communications, and after 1933 his colleague and friend Wong Wen-hao was chairman of a government commission on natural resources for defense, for which Ting helped plan a number of industrial projects to be put into operation in the interior in case of war. He chaired a preparation committee for the establishment of an iron and steel plant at Ma-an-shang, and the last work he undertook before his death was a survey of mines and mineral resources in southern Hunan along the proposed route of the Hankow-Canton railway, together with a study of the facilities which the region around Changsha might offer to academic institutions if they were forced to evacuate the north.[43] But he refused several opportunities for full-time official posts, probably connected with planning for heavy industry,[44] and instead spent what proved to be the last two years of his life absorbed in work which gave particular scope to his own combination of academic and administrative competence. In the summer of 1934 he became secretary-general (*tsung-kan-shih*) and thus effective executive head of the Academia Sinica (*Chung-yang yen-chiu-yüan*), the government-sponsored institute for conducting and coordinating academic research in China.

When Ting came to the Academia Sinica it consisted of a federation of individual research organs, under the presidency of Ts'ai Yüan-p'ei, which divided their energies between independent research and the provision of scientific data for the government. As secretary-general, Ting was an ambitious rationalizer of the Academia Sinica's administration. His goal was the formation of an integrated system of research organizations which would operate out of the universities, the government, and the

Academia Sinica itself, but whose projects would be co-ordinated by the academy's management on the basis of importance and relevance to national need. Such an organization would have at its disposal the nation's reservoir of scientific expertise, placed at the service of a rationalized set of development priorities. It was a grandiose conception: the academic intelligentsia mobilized in the service of the state's technocratic goals. Although he respected the limits posed by the Academia Sinica's structure, which meant that no one could actually presume to dictate projects to independent research agencies, Ting's reforms necessarily involved a considerable concentration of power in the Academia Sinica executive, and the reforms did not fail to ruffle some academic feathers.[45]

As a scientific positivist Ting did not believe that there existed a categorical distinction between "pure" and "applied" scientific research. He was not ashamed to have primarily utilitarian ambitions for the Academia Sinica. "The Academia Sinica's most important, most practical duty is profitably to use the scientific method to investigate our raw materials and production, in order to solve all kinds of industrial problems," he said.[46] He expressed particular interest in research into natural resources and into new uses and markets for resources abundantly available already, in the study of nutrition and of mass communications. Although to the public he defended the importance of pure research, it was by stressing its ultimate usefulness, and he considered it a relic of the old classicism to honor the pure science of the academies at the expense of its applications to practical life. He stated frankly that a poor and backward country did not have the resources to lead the way in new, fundamental discoveries.[47] Under Ting's leadership the Academia Sinica anticipated its later, more thoroughgoing, centralization at the hands of the Chinese Communists, and its canalization, like all Chinese science, into projects directly relevant to national goals. Ting himself was a counterexample to the Communists' later claim that Republican scientists had preferred the theoretical problems of the laboratory over those applicable to national development.

Ting's death in January 1936, just three months before his forty-ninth birthday, was sudden and premature enough to shock everyone but perhaps himself. Several times he had remarked fatalistically to friends that he would have to accomplish his life's work by fifty because the men in his family did not live past that age. In 1933 Ting had returned from Russia badly overtired and complaining of numbness in his extremities; subsequently doctors at Peking Union Hospital diagnosed hardening of the arteries. For a time he was depressed over his condition and ready even to talk of retiring to a life of quiet research.[48] But gradually he recovered spirits and plunged into the work of the Academia Sinica with his usual formidable energy. Aside from giving up smoking, he in no way changed his regimen, nor his accustomed habits of vigorous exercise, and he told Fu Ssu-nien that since nothing could be done about his health, he proposed to ignore it.

In early December 1935 he arrived in Hunan and embarked on a strenuous program of mineral prospecting for the government. After a week of work, which involved several days of arduous, mountain hiking, he was found one morning in a coma in his room at a small inn near Hengyang. For over three weeks he lay ill in the hospital, first at Hengyang and then at Changsha, victim of a series of medical bungles which both concealed and exacerbated his true condition. Believing that the original coma had been caused by poisoning from a poorly ventilated coal stove, the doctors at first applied artificial respiration, and in so doing broke one of Ting's ribs. A serious infection of his pleural cavity developed as a result and, while he was still undergoing treatment for the latter condition, he lapsed into a second coma and, on January 5, 1936, died. The circumstances of his illness led many people, then and afterward, to blame his death on an accident, compounded by the ineptitude of doctors whose Western training appeared disillusioningly ineffectual in this emergency. However, an autopsy disclosed that Ting had also suffered one or more cerebral hemorrhages, a calamity which his overall medical history had made a possibility for several years.

In accordance with the request in his will that he be buried in the place of his death, Ting's body was interred near Changsha. As a boy about to leave for England, the emotions of that occasion had led young Ting to imagine his future death in a mood of traditional, yet touching, bravado. In a poem composed for a relative, he had rejected a Confucian burial custom:

> The bold lad goes out from his village gate —
> He vows to succeed or never come back.
> Who needs the trees of home by his grave?
> Among men the world over, hills are green.[49]

Though the instruction never changed, its implications did. While it still has meaning, a traditional practice can be self-consciously rejected — and the Confucian exile's grave was in fact an antipode which helped give the Confucian family temple its potency. But what Ting rejected as a boy as an adult he simply ignored, and when he was buried he lay in Changsha not out of inverse ancestral piety, but in simple indifference to anachronistic family customs.

Chapter IX

The Rationalist Between East and West

Ting Wen-chiang probably knew and understood European culture, particularly that of the Anglo-Saxon world, as well as any Chinese of his generation. But although his intellectual development was an unbroken evolution away from Confucianism and toward the stance of an international twentieth-century intellectual, this is not to say that the process was a simple "Westernization." His development away from Confucianism indeed was irresistible, brought about by the intellectual influence of Western learning and by the post-imperial Chinese reality which no longer sustained Confucian patterns of individual action. A good scientific education quickly destroyed in Ting all remnants of the Confucian view of nature as a harmoniously interacting cosmos linked inextricably with the social purposes of man. The analytic methods of scientific positivism supplied for him radically untraditional standards for determining the truth about virtually all matters of substance. The experience of fighting for political power in the Republican environment more gradually undermined his original image of political life and of the relation between knowledge and action. From a belief in rule by virtue and confidence that political action was the natural outgrowth of scholarship and the duty of scholars, Ting gradually changed to a more specialized view of political leadership which stressed the scientific expertise of any ruling elite and envisaged the possible alienation of the pure intellectual from the political process. However, in all these transformations, Ting displayed a consistent gravitation toward that Western content which could be grafted onto Chinese structures.

Social Darwinism was particularly adaptable to the frame of reference of a Confucian-bred Chinese. Ting saw in Darwinism a scientific explanation for the existence of the kind of social structure in which he had been brought up: one which established a hierarchical order of authority, yet tempered this authority with an ideal of merit in power, and the hope, if not the expectation, for social self-sacrifice and altruism on the part of those who wielded power. Genetics taught the existence of a group of naturally superior men, nature's candidates for leadership; and natural selection, operating in the context of an entire human society, seemed to condition the masses of people to further group welfare in a way easily illustrated by the corporate family and clan customs of rural China. Moreover, by interpreting certain human social and ethical ideals as facts of nature, this version of social Darwinism preserved in attenuated form the link between natural processes and human values which had been an essential feature of the Confucian image of the universe.

In spite of Britain's democratic institutions, the social atmosphere of the England Ting knew remained tenaciously stratified, and, in order to select the elite, Englishmen displayed a flexible combination of reliance upon birth — gentility — and confidence in education and self-improvement, which accorded neatly with the self-image of the Chinese gentry in its own setting. Thus Ting was able to accept quite easily certain social assumptions underlying the writings of Victorian scientists like Huxley,

Galton, and Pearson — assumptions that whatever the participation of the populace in the rituals of self-government or in the benefits of industrial abundance, the leadership, both intellectual and practical, lay with an upper class formed by organic social processes of differentiation. Ting's intellectual identification with late nineteenth-century British scientists confirmed his native sense of a natural class structure in society, and it left him permanently skeptical of the merits of egalitarianism. Neither the prospect of mass education nor that of economic leveling, both of which he accepted as among the goals of modernization in China, in any fundamental way changed his view; and he remained a person whose intellectual commitments as well as his personal social preferences tied him to that small group of gentry-bred intelligentsia which remained remote from the mass movements that were becoming the actual agents of political change.

Further, in an age when China was learning to accept the fact that it shared the globe with other civilizations as venerable and, for the time being, much more powerful than itself, the Darwinian world view embodied a historicism ample enough to integrate Chinese history with the newly discovered history of the world and generous enough to minimize the importance of China's present relative backwardness in material culture. As a Chinese he thought naturally in millennia, and it was comforting for him to consider his country's contemporary crisis not merely against a backdrop of dynastic cycle, but even more broadly in the perspective of mankind's long biological struggle to create civilization and material abundance. It was a short step from a view of global geological and biological history as a single development to one which sought instances of the interrelatedness of cultures — East and West — during recent civilized ages; the findings of modern sociology, anthropology, and history suggested to the Chinese Darwinian confirming instances of such interaction. The history of China, if no longer the only history there was, now became part of world history and on a basis of equality.

Moreover, the Darwinian scientists offered an empirical account of the scientific method. The first generation of Chinese students of Western science were in general attracted to empirical and historical disciplines: medicine, geology, biology, archaeology. Relatively few of them became physicists or chemists. Both in practice and in accounts of how that practice operated, the natural history sciences of the nineteenth century were more immediately accessible to a Chinese whose forebears had considered historical studies the model for all rational inquiry, and who was not accustomed to distinguish between theoretical concepts in science and speculative thought in general. So Ting found in the natural history sciences in England a characterization of scientific methodology which seemed both intuitively illuminating and legitimately applicable to the activities of certain traditional scholars in China, as well as to his own. In this way a favorable rationale was created for that association of Western science with traditional Chinese scholarship which was an important feature of his generation's efforts at Sino-Western cultural integration. He was able to feel a kinship with figures like Sung Ying-hsing and Hsü Hsia-k'o, and to lend the prestige of the name, science, to their pursuits. Their inquiries did, in fact, like his own, fit a concept of scientific thought which stresses classification and inductive generalization rather than the more fertile and fundamental techniques of theoretical hypothesis and mathematical model building.

To talk of relations between Darwinian scientific content and traditional Chinese patterns is not, however, to suggest that Ting's intellectual make-up was simply predetermined by inherited predispositions. Anglo-Saxon positivism and social Darwinism became staples of his outlook partly because they were there: dominant intellectual currents in the Edwardian England where he was educated. If these doctrines contributed to the twentieth-century transformation of the Chinese consciousness, they were also in the independent process of evolving within the Western world and so stimulating ongoing modification in Chinese responses. The temporal perspective is important, not only for what Ting brought to his Western education, but for what he found there.

In addition, if the structure of Ting's adaptations cannot be explained except with reference to the historical possibilities of his intellectual experience, both East and West, neither can it be described only in terms of a set of psychological imperatives. A belief can be adopted solely because it satisfies some need, and the forms taken by many of the doctrines professed by early twentieth-century "Westernized" Chinese have frequently been attributed to the Asian's longing to preserve his faith in the value of Chinese culture and the justifiability of his attachment to it. At the same time, he adopts Western ideas which impose themselves by the prestige of their success. The importance of this insight may be assessed by the fact that it has almost assumed the status of a cliché. But though historians are instinctive sociologists of knowledge, who generally tend to explain ideas in terms of the social (or psychological) environment in which they appear, this can be a dangerous path, particularly when followed with reference to intellectual convictions arrived at by a process of reasoning for which thinkers make claims to autonomy. Ting was nothing if not an intellectual, and he had the intellectual's special belief in the efficacy of rational thought and in the obligation of honest thinkers to follow the argument where it leads, no matter what the consequences. His positivism — that is, his conviction that everything human beings may reliably accept as knowledge must be arrived at by scientific procedures of thought — was the cornerstone of his intellectual creed.

Positivism both implied a confidence that some conclusions may simply be accepted as correct and suggested an impartial method by which such correct conclusions may be arrived at. In daring to say that his convictions, selected from the variety of intellectual possibilities open to the student of both Europe and China, might simply be *right*, he made a claim for the autonomy and integrity of intellectual inquiry which cannot be ignored, and which supplied an ideal higher and more universal than that of cultural allegiance to either Europe or China. In the light of that method and that ideal, the injunction "select the best from East and West" provided more than a formula whereby a culturally displaced person might integrate his experience; it also suggested a guide to what any rational person must accept. It would be wrong to say that Ting's convictions were the result only of rationalization, never of reasoning.

However, it should be possible to say that Ting was a rationalist without also implying the impossible ideal — that he thought in an a-historical, purely logical fashion upon all occasions. Fundamentally, rationalism is not a matter of thought itself but of attitude toward thought: it implies confidence in the capability of reason to deal with the perplexities of life and commitment to the possibility of a rational methodology. The limitations of Ting's rationalism were most evident in his political career. In practical politics the positivist was committed to the scientific analysis of issues involving human social welfare. In practice in China this commitment led to social action which was no less ineffectual for its professed "pragmatism." Once again, as in the case of Ting's philosophy of science, the Chinese was led astray by doctrines which had been formulated in an uncertain manner by his European teachers themselves. The idea of a scientific social revolution was one of the staples of positivist propaganda in early twentieth-century Europe, but men like Karl Pearson and John Dewey promoted it without supplying for it a method capable of resolving the fundamental problem of whether the model of the scientific experiment can ever be successfully adapted to the process whereby men, acting as both self and object, observer and observed, analyze their own consciousness and experience. In the Anglo-Saxon world, where basic political institutions were comparatively stable, some achievements proved possible without successful self-analysis. Pragmatists could suggest fairly concrete individual reforms, such as Dewey's educational proposals or Pearson's genetic ones — proposals which might lead to improvement in the quality of life in a society, but which would operate by modifying and adjusting existing institutional patterns, not by wholly revolutionizing them. But pragmatism's plausibility in fundamentally conservative English-speaking nations could not be matched in China, where a truly radical social upheaval was underway.

Having no really solid methodology, only the pretension to one, pragmatists in China were reduced to rejecting ideology as obviously unscientific, and exhorting instead, that experimental tests of "practical" policies be carried on, just as scientific theories are tested in the laboratory. In fact, pragmatism in China permitted the substitution of a hidden, implicit, conservative ideology, tacitly assumed but not consciously formulated by its adherents, in place of an open, dogmatically adopted, radical ideology. For Ting and his fellow nonparty intellectuals, those policies which seemed practical were exactly those which could be carried out by introducing new content into the structure of an authoritarian bureaucratic government — that is, within a framework which had been provided by the former monarchy but which in Republican China no longer really existed. His scientific pragmatism did not disturb Ting's essentially conservative idea of government as properly the business of an autocratic leadership operating in collaboration with a specially trained and selected bureaucratic elite — an idea which in its apotheosis of "good men" retained much of the paternalistic, scholarly, and ethical emphasis of traditional Confucian statecraft. It led him and other intellectual partisans of scientific political action to turn to the forces which most nearly reproduced the political structure of the prerevolutionary empire, the warlords, and to attempt in a period of disunion and chaos to create an up-to-date version of the administrative apparatus which had traditionally been at the center of the political process. In the "good government" experiment and as director-general of Greater Shanghai, Ting acted upon these assumptions and failed.

In the revolutionary interregnum of Republican China, actual political change was effected by forces which had no respected place in the Confucian model of the state — by armies, or by the masses of the people, who were now being taught to add to the traditional force of peasant revolt the ideological inspiration of acquired Western doctrines of nationalism and socialism. Ting could not help being aware of these new political forces, but his efforts to harness them to his own political goals proved ineffectual.

He did participate in attempts to create a mass public opinion which could agitate for reforms as the May Fourth demonstrators had fought for national sovereignty. But these attempts failed when they were carried out through the medium of coterie journalism by gentry-oriented intellectuals whose social customs and concept of the role of the intellectual kept them at a distance from the illiterate populace. On the other hand, efforts to cooperate with warlords failed when it became apparent that a bureaucratic entourage would not succeed in circumscribing a militarist engaged in civil war with those checks and balances ideally imposed on a ruler by a Confucian administrative apparatus. Because of these inherited patterns parading under the label of pragmatism, Ting's political career appears in retrospect not merely unlucky, but misconceived, in spite of the fact that the practical policies which he advocated in matters like taxation, economic development, and diplomacy were models of modernity and good sense.

In the 1930s, however, Ting did not repeat his political errors of the 1920s. Even under the superficially more favorable conditions provided by the Nationalist government, he resisted the temptation to be drawn again into a militarist's political apparatus. The *Independent Critic* had less strident political pretensions than did *Endeavor,* and it operated more resignedly as the voice of an academic clique. Moreover, the long-drawn-out crisis, exacerbated by Japan's invasion of the mainland, made clearer to him the inadequacy of pragmatic political solutions when offered in an environment of deep ideological and factional strife. He no longer looked to purely scientific solutions for the problems of society, but to the once scorned idea of ideology, a common belief to enable people to work together for needed goals. His own idea of the structure of government continued to focus upon the authoritarian, bureaucratic structure suggested by China's past, but at the end of his life he made one further attempt to update its content. He advocated a technocratic, socialist despotism which he modeled quite closely on his interpretation of Stalin's Russia.

In his political as well as in his social and scientific thought, Ting's adaptations from England remained selective. Here again it is possible to see the interaction of Ting's rationalism and the historical possibilities available to both England and China, producing a set of political beliefs which remained almost untouched by Anglo-Saxon political liberalism. In England the Fabian socialists did represent a crosscurrent of British democratic thought which paid less overt attention to the individual as the fundamental unit of society than to forms of state organization which might guarantee an equality threatened by competitive individualism in practice. Among Westerners, Ting preferred Fabians like Harold Laski and H. G. Wells, particularly when they echoed his own doubts about the democratic process and his preoccupation with state-generated administratively oriented social engineering. In England this group also came closest to sharing his own attraction to Soviet Communism, but Ting probably felt the deeper pull. Russia's authoritarian, bureaucratic state structure was closer to Chinese precedent than to the heritage of parliamentary democracies. But equally important, generations of Chinese state leadership by an intellectual aristocracy — schooled to consider political power as the companion of humanistic culture — had left Ting innocent of the liberal Englishman's and American's suspicion that the power of the state inevitably tends to threaten individual rights and values. In traditional China the successful exercise of state power was deemed due to the harmony of rulers with fundamental principles; contemporary authoritarian government, in Ting's mind, would succeed when leaders learned to respect those modern principles of action which were dictated by scientific research. Where facts (or principles) must dictate choices to rational (or good) men, the free act of making choices must seem of secondary importance. In this way liberalism — with its reverence for individual dissent and its reliance upon a consensus which is merely the product of majority wish — becomes at least irrelevant and at worst dangerously relativistic.

If his kind of rationalism helped keep Ting from being a liberal, it also kept him from being a revolutionary. In spite of his lifelong interest in military affairs, instinctively Ting always looked to the operations of human intelligence to provide the solution to problems; orderly procedures, in life and government, struck him as unquestionably good. Persistently he relied upon the effectiveness of the rationalist, technical solutions which the intellectual habit made him conjure up so readily. The Communists in China, professing to operate from intellectual premises (the method he respected). disappointed him by selecting as those premises doctrines he rejected as intellectually mistaken; further, the Communists alienated him by linking their doctrines to incitement to the revolutionary revolt, which he saw as irremediably destructive of all orderly planning and construction. Perhaps it is one of the intellectual's perennial problems that he tends to assume that the ratiocinative procedures which have exercised such control over his own existence, both in guiding his conscious choices and as the monitor of his own chaotic instincts, are in fact capable of extending their domain over the external world at large. Understanding becomes its own goal, an insidious form of "magical thinking" for the reasoner himself. Dazzled by the logical ir-refutability of seen truth, the intellectual has an illusion of mastery over things and events analyzed. If this is so, in a revolutionary era a rationalist like Ting may indeed be disarmed.

Appendixes / Notes
Bibliography / Glossary / Index

Appendix A

CHRONOLOGICAL LIST OF THE PRINCIPAL EVENTS IN TING'S LIFE

April 13, 1887	Birth
1898	The Hundred Days' Reform
Autumn 1902	Ting leaves for Japan as a student
February 1904	Outbreak of the Russo-Japanese War
Summer 1904	Ting sails for England
1904–06	Attends secondary school in Spaulding, Lincolnshire
Fall 1906	Spends one term at Cambridge University
Winter 1906–07	Travels on the Continent
1907–08	Attends a Glasgow technical college
1908–11	Attends and graduates from Glasgow University
May–July 1911	Ting returns home to China, traveling via Yunnan and Kwei-chow provinces
October 10, 1911	Outbreak of the Wuchang revolution
1911–12	Ting teaches at the *Nan-yang chung-hsüeh* in Shanghai
February 1913	Joins the Bureau of Mines in Peking. The Geological Research Institute is organized
Nov.–Dec. 1913	Field trip to Shansi with Solger
January 1914	Ting's father dies
February 1914–Jan. 1915	Field trip to Yunnan

241

1916	Geological Survey is founded with Ting as director
Spring 1917	Ts'ai Yüan-p'ei becomes chancellor of Peking University
February 1919–	Ting travels in Europe with Liang Ch'i-ch'ao
May 4, 1919–	Demonstrations in Peking against the Versailles treaty
August 1920	Fall of Tuan Ch'i-jui's government in Peking
1921	Ting resigns from the Geological Survey to become managing director of the Pei-p'iao mine in Jehol
January–May 1922	First Fengtien-Chihli war, leading to Wu P'ei-fu's call for a constitutional restoration
May 13, 1922	*Endeavor* manifesto
September 19, 1922	*Endeavor* members join the "good government" cabinet
November 25, 1922	Fall of the "good government" cabinet
Feb.–Dec. 1923	Science and metaphysics debate
October 10, 1923	Ts'ao K'un elected president of the republic
August–Oct. 1924	War between Kiangsu and Chekiang
May 30, 1925–	Anti-imperialist demonstrations in Shanghai and other cities
Winter 1925	Ting resigns from Pei-p'iao mine
February–May 1926	Serves on the Sino-British Commission for the rendition of the Boxer indemnity
May 5, 1926	General Sun Ch'uan-fang inaugurates the Greater Shanghai municipality with Ting as director-general
June 1926	The Northern Expedition is launched by the Kuomintang
December 31, 1926	Ting formally resigns as director-

	general of Greater Shanghai
March 1927	Kuomintang armies reach Nanking and Shanghai
1927	Ting lives in retirement in Peking and Dairen
March–Dec. 1928	Field trip to Kwangsi
December 1928	Ting returns to Peking and the Geological Survey
Spring 1929– Summer 1930	Ting takes charge of a major field trip to the southwestern provinces
September 19, 1931	Japanese seizure of Manchuria
October 1931	Chiang Meng-lin becomes chancellor of Peking University and Ting becomes research professor of geology
February 1933	Japanese conquer Jehol
May 31, 1933	Tangku truce
June 1933	Ting leaves Shanghai for the International Geological Congress, Washington, D. C. and Europe
Sept.–Oct. 1933	Travels in the Soviet Union
June 1934	Becomes secretary-general of the Academia Sinica
December 2, 1935	Arrives in Changsha for coal-prospecting expedition
December 9, 1935	Taken ill in Hengyang, Hunan
January 5, 1936	Death

CHRONOLOGICAL LIST OF TING WEN-CHIANG'S TECHNICAL PUBLICATIONS

"T'iao-ch'a Cheng-t'ai t'ieh-lu fu-chin ti-chih k'uang-wu pao-kao shu" 調查正太鐵路附近地質鑛務報告書, 農商公報 (Report of an investigation of the geology and mineral resources of the region adjoining the Cheng-t'ai Railroad); *Nung shang kung-pao* (Agriculture and commercial gazette), Vol. 1, Nos. 1 and 2 (1914). Published with Solger and Wong.

"Chihli Shansi chien Wei-hsien Kwang-ling Yang-yüan mei-t'ien pao-kao" 直隸山西間蔚縣廣靈陽原煤田報告, 地質彙報 (Report on the coal fields of Wei-hsien, Kwangling and Yangyüan of Chihli and Shansi Provinces); *Ti-chih hui-pao*, No. 1 (1919).

"Ching-ch'ao ch'ang-ping hsien hsi-hu tsün meng k'uang" 京兆昌平縣西湖村錳鑛, 地質彙報 (The manganese deposits of Hsi-hu village, Chang-ping and Ching-ch'ao hsien, Chihli); *Ti-chih hui-pao*, No. 1 (1919).

"Report on the Geology of the Yangtze Valley Below Wuhu," Whangpoo Conservancy Board, Shanghai Harbor Investigation Ser. 7, Reprint, Vol. 1 (1919). A Chinese translation prepared by Wang Hu-chen appeared in the *Yangtze chiang shui-tao cheng-li wei-yüan-hui yüeh-k'an* 揚子江水道整理委員會月刊 汪胡楨 (Bulletin of the commission for the reclamation of the Yangtze waterway), Shanghai, Vol. 1, Nos. 1, 2, and 3 (1919).

K'uang cheng-kuan chien, fu hsiu-kai k'uang-yeh t'iao-li i-chien shu 鑛政管見附修改鑛業條例意見書 (Mining management, with appended opinions on the reform of mining industry regulations), Geological Survey Pamphlet, 1920. Published with Wong Wen-hao.

"Pei-ching ma-lu shih-liao chih yen-chiu" 北京馬路石料之研究, 農商公報 (Study of stone materials for Peking highways); *Nung shang kung-pao*, Vol. 7, No. 11 (June 1921).

"Yangtze chiang hsia-yu tsui chin chih p'ien-ch'ien—san chiang wen-t'i" 揚子江下游最近之變遷　三江問題，北京大學地質研究會會刊 (The problem of the three rivers: Most recent shifts in the waters of the lower Yangtze River); *Peiching ta-hsüeh ti-chih yen-chiu-hui hui-k'an* (Bulletin of the Peking University Geological Research Society), No. 1 (October 1921).

"The Tectonic Geology of Eastern Yunnan," *Congress geologique internationale, 13éme session belgique, comptes rendus,* fasc. 2, 1922.

The sections on geology in J. G. Andersson's *The Cenozoic of North China,* published in the Mem. East. Survey, China, Ser. A, No. 3 (1923).

"Note on the Gigantopteris Coal Series of Yunnan," published in A. W. Grabau's *Stratigraphy of China,* Pt. 1, 1923.

"Ti-chih hsüeh hsü" 地質學序 (Preface to "Study of Geology"), published in *Ti-chih hsüeh* by C. Y. Hsieh, Shanghai, 1924.

"On the Nephelite Syenite of Maokou in Hweili District Szechuan" (abstract), *Bulletin of the Geological Society of China,* Vol. 4, No. 1 (1925). With W. H. Wong.

"Stratigraphical Note," in T. G. Halle, "Fossil Plants from Southwestern China," *Paleontologica Sinica,* Ser. A, Vol. 1, fasc. 2, pp. 22–24 (1927).

"The Triassic System—China Proper," in A. W. Grabau, *Stratigraphy of China,* pp. 37–52 (1928).

"The Orogenic Movements in China," *Bulletin of the Geological Society of China,* Vol. 8, No. 2 (July 1929).

"Notes on the Language of the Chuang in North Kwangsi," *Bulletin of the Museum of Far Eastern Antiquities,* No. 1 (1929).

"Kwangsi chuang yü chih yen-chiu" 廣西獞語之研究，科學 (Research into the language of the Chuang of Kwangsi); *K'o-hsüeh,* Vol. 14, No. 1 (1929).

"On the Stratigraphy of the Fengninian System," *Bulletin of the Geological Society of China,* Vol. 10 (1931).

"A Statistical Study of the Difference Between the Width-height Ratio of Spirifer Tingi and That of Spirifer Hsiehi," *Bulletin of the Geological Society of China,* Vol 11, No. 4 (1932).

"The Permian of China and Its Bearing on Permian Classification," *Reports of the Sixteenth International Geological Congress,* Washington, D. C., 1933. Reprint, 1934. Published with A. W. Grabau.

"The Carboniferous of China and Its Bearing on the Classification of the Mississippian and Pennsylvanian," *Reports of the Sixteenth International Geological Congress,* Washington, D. C., 1933. Reprint, 1934.

Published with A. W. Grabau.

"The Mississippian and Pennsylvanian of China and their bearing on the classification of these systems," (abstract), *Pan-American Geology*, Vol. 60, pp. 232–233 (1933). With A. W. Grabau.

"Notes on the Records of Droughts and Floods in Shensi and the Supposed Desiccation of Northwest China," *Geografiska Annaler*, Vol. 14, honoring the seventieth birthday of Sven Hedin, 1935. Published with W. H. Wong. A Chinese translation appears in *Fang-chih yüeh-k'an* 方志月刊, Vol. 9, No. 2 (April 1936).

"On the Influence of the Observational Error in Measuring Stature, Span and Sitting-height Upon the Resulting Indices," published in *Ching-chia Ts'ai Yüan-p'ei hsien-sheng* 慶賀蔡元培先生 (In honor of Mr. Ts'ai Yüan-p'ei), 1935.

"Cambrian and Silurian Formations of Malung and Chütsing Districts, Yunnan," *Bulletin of the Geological Society of China*, Vol. 16 (1936–1937). Published with Y. L. Wang and posthumously edited by T. H. Yin.

"Yunnan Ko-chiu fu-chin ti-chih k'uang-wu pao-kao" 雲南箇舊附近地質鑛物報告 (Report on the mineral deposits in the region around Kokiu in Yunnan); *Ti-chih chuan-pao* 地質專報, Ser. B, No. 10 (1937). Published posthumously, edited by T. H. Yin.

Note: Some entries on this list are based upon two earlier bibliographies of Ting's writings: I. W. Shen, "List of Writings of Dr. V. K. Ting on Geology and Allied Subjects," in *Geological Reports of Dr. V. K. Ting* (Nanking, 1947); and Chang Ch'i-yün 張其昀, "Ting Wen-chiang hsien-sheng chu-tso hsi-nien mu-lu 丁文江先生著作繫年目錄"; *Tu-li p'ing-lun*, No. 188 (Feb. 16, 1936).

NOTES

ABBREVIATIONS USED IN THE NOTES

BGSC *Bulletin of the Geological Society of China*
CYYCY *Chung-yang yen-chiu yüan yüan-k'an*
TLPL *Tu-li p'ing-lun*
KHY *K'o-hsüeh yü jen-sheng kuan*

1. INTRODUCTION

1. Y. C. J. Wang, *Chinese Intellectuals and the West* (Chapel Hill, N. C., 1966). See especially pp. 156–164 and 168–187. See also Mary Wright, "Pre-Revolutionary Intellectuals of China and Russia," *China Quarterly*, No. 6: 175–179 (April-June, 1961).

2. Liang Ch'i-ch'ao, "Hsin min shuo" (A new people), *Yin-ping-shih ho chi*, Vol. 19.

3. Joseph Needham, *Science and Civilization in China*, Vol. 2 (Cambridge, England, 1956). See especially pp. 279–293.

4. Joseph R. Levenson, *Confucian China and Its Modern Fate*, Vol. 1 (Berkeley, 1958).

5. See especially Knight Biggerstaff, *The Earliest Modern Govern-*

ment Schools in China (Ithaca, 1961), Chaps. 3 and 4.

6. T'an Ssu-t'ung, *Jen hsüeh* (Peking, 1958), pp. 2–6.

7. In the "science and metaphysics debate" of 1923, such sentiments were still described as "commonplace." See p. 2 of Ting Wen-chiang's "Hsüan-hsüeh yü k'o-hsüeh ti t'ao-lun ti yü-hsing" (Envoi to the debate on science and metaphysics), in *KHY*.

2. THE RETURNED STUDENT

1. In adopting an English version of his name, Ting romanized initials according to the pronunciation of his native Shanghai dialect. He also used the Chinese courtesy name of Ting Tsai-chün, and occasionally he wrote under the pen name of Tsung Yen, adopted in honor of his fellow provincial Fan Chung-yen.

2. The most important sources for Ting's personal history are the issue of the *Tu-li p'ing-lun* (Independent critic) devoted to articles by Ting's friends and associates on the occasion of his death (No. 188, Feb. 16, 1936); and the biography of Ting by his close friend Hu Shih, which was published in a special issue of the *Chung-yang yen-chiu yüan yüan-k'an* (Annals of the Academia Sinica) commemorating the twentieth anniversary of his death (No. 3, December 1956). Other recollections of Ting by those who knew him, plus reprints of a number of his non-political writings, also appear in this volume.

3. See Ting Wen-t'ao, "Wang ti Tsai-chün t'ung-nien i-shih chui-i lu" (Recollected anecdotes of the youth of my lost brother Tsai-chün), *TLPL*, No. 188: 45–48 (Feb. 16, 1936).

4. Ting Wen-chiang, comp., *Liang Jen-kung hsien-sheng nien-p'u chang-pien ch'u-kao* (Draft materials for a chronological biography of Mr. Liang Jen-kung; Taipei, 1958). Preface by Ting Wen-yüan, p. 5.

5. Ting Wen-t'ao, *TLPL*, No. 188: 46 (Feb. 16, 1936).

6. Hu Shih, *Ting Wen-chiang ti chuan chi*, p. 5.

7. Ting Wen-chiang, "Hsien-tsai chung-kuo ti chung-nien yü ch'ing-nien" (The middle-aged and the young in China today), *TLPL*, No. 144: 8–9 (Mar. 31, 1935).

8. V. K. Ting, "Chinese Students," *Westminster Review*, 169.1: 48–55 (January 1908). This article is a description for a European audience of Chinese student life in Japan.

9. Li Tsu-hung (Li I-shih), "Liu-hsüeh shih-tai ti Ting Tsai-chün" (Ting Tsai-chün as a student abroad), *TLPL*, No. 208: 12–18 (July 5, 1936). See also Ting Wen-chiang, *Liang Jen-kung hsien-sheng nien-p'u*, preface by Ting Wen-yüan, p. 5.

10. T'ang Ai-li (T'ang Chung), "Tui-yü Ting Tsai-chün hsien-sheng ti hui-i" (Recollections of Mr. Ting Tsai-chün), *TLPL*, No. 211: 15 (July 26, 1936).

11. Li Tsu-hung, "Liu-hsüeh shih-tai ti Ting Tsai-chün," p. 13.

12. Ting Wen-chiang, "Su-o lü-hsing chi" (Record of a journey to the Soviet Union), *CYYCY*, 3: 451–453 (December 1956).

13. Ting Wen-chiang, "Hsüan-hsüeh yü k'o-hsüeh: Ta Chang Chün-mai" (Metaphysics and science: A reply to Chang Chün-mai); in *KHY*, pp. 23–24.

14. Hu Shih, "Ting Tsai-chün che-ko jen" (This man, Ting Tsai-chün), *TLPL*, No. 188: 9.

15. Fu Ssu-nien (Fu Meng-chen), "Ting Wen-chiang i-ko jen-wu ti chi pien kuang-ts'ai" (A few of the glories of Ting Wen-chiang as a personality), *TLPL*, No. 189: 9 (Feb. 23, 1936).

16. Pierre Teilhard de Chardin, *Letters From a Traveller*, Bernard Wall (London, 1962), p. 126.

3. ODYSSEY OF A CHINESE SCIENTIST

1. This trip, like others, was described in "Miscellaneous Travels" (*Man yu san chi*), a series of informal essays which Ting wrote between 1932 and 1936 for *Tu-li p'ing-lun*. They are reprinted in *CYYCY*, No. 3: 341–437.

2. Fu Ssu-nien (Fu Meng-chen), "Wo so jen-shih ti Ting Wen-chiang hsien-sheng" (The Ting Wen-chiang I knew), *TLPL*, No. 188: 3 (Feb. 16, 1936).

3. Ting Wen-chiang, comp., *Liang Jen-kung hsien-sheng nien-p'u*, preface by Ting Wen-yüan, pp. 5–6.

4. Reorganizations in 1913 placed the geology section under the Ministry of Agriculture and Commerce (*Nung shang pu*). See Hu Shih, *Ting Wen-chiang ti chuan chi*, p. 17.

5. For Ting's account of defects in Chinese cartography, written in reply to a letter to the editor, see *TLPL*, No. 19: 20–22 (Sept. 25, 1932).

6. Ting Wen-chiang, "Man yu san chi," pp. 345–346.

7. The textbook in question, *Tung-wu-hsüeh chiao-k'o-shu*, was published in Shanghai in 1914. See Li Hsüeh-ch'ing, "Chiu-nien Ting shih Tsai-chün hsien-sheng" (Recollections of Professor Ting Tsai-chün; in *Ti-chih lun-p'ing: Ting Wen-chiang hsien-sheng chi-nien hao* (Geological discussions: issue in memory of Mr. Ting Wen-chiang), 1.3: 237 (July, 1936).

8. *Ti-chih hui-pao* (Bulletin of the Geological Survey of China),

Vol. 1, No. 1 (July 1919), Chinese preface by Ting Wen-chiang.

9. W. H. Wong (Wong Wen-hao), "Richthofen and Geological Work in China," *BGSC*, 12.3: 311–313 (1933).

10. Similarly, even later, to Hollington Tong, Ting's geological studies were "like the *Shan hai ching*." See Tung Hsien-kuang, "Wo ho Tsai-chün" (Tsai-chün and I), *CYYCY*, No. 3: 135.

11. *Ti-chih hui-pao*, Vol. 1, No. 1, Chinese preface.

12. Ferdinand von Richthofen, *China: Ergebnisse einer Reisen und darauf gegründeter Studien*, Berlin, 1877–1912. See I, xxxviii.

13. *Ti-chih hui-pao*, Vol. 1, No. 1, English preface by V. K. Ting. Ting omitted the reference to avarice.

14. Ting Wen-chiang, "Man yu san chi," pp. 360–361.

15. This class included many of the most prominent Chinese geologists of the next twenty years, in particular Hsieh Chia-yung, Wang Chu-ch'üan, Yeh Liang-fu, Li Chieh, T'an Hsi-ch'ou, Chu T'ing-hu, and Li Hsüeh-ch'ing. See Amadeus Grabau (*Ko-li-p'u*), "Ting Wen-chiang hsien-sheng yü Chung-kuo k'o-hsüeh chih fa-chan" (Mr. Ting Wen-chiang and the development of Chinese science) translated into Chinese by Kao Chen-hsi; *TLPL*, No. 188: 20–23 (Feb. 16, 1936).

16. Ting Wen-chiang, "Man yu san chi." See pp. 361–369 for Ting's account of his trip to Shansi, and pp. 373–430 for his trip to Yunnan. Ting's full, technical field notes in English from these and other trips are printed in *Ting Wen-chiang hsien-sheng ti-chih t'iao-ch'a pao-kao* (Geological Reports of Dr. V. K. Ting) published by the Geological Survey (Nanking, 1947).

17. V. K. Ting, "On Hsü Hsia-k'o (1586–1641), Explorer and Geographer," *New China Review*, 3.5: 325–337 (October 1921).

18. Huang Chi-ch'ing, "Ting Tsai-chün hsien-sheng tsai ti-chih hsüeh shang ti kung-tso" (Mr. Ting Tsai-chün's work in geology); *TLPL*, No. 188: 23 (Feb. 16, 1936).

19. Ting stayed in Kokiu for about two months, but partly because of a temporary illness he did not complete his mineral survey of the district. Many years later he organized some of his students for a follow-up expedition, and, with the help of his maps, they were able to begin a comprehensive analysis of the deposits. They published the results under Ting's name after his death. See Ting Wen-chiang, "Man yu san chi," pp. 373–387.

20. These portions of Ting's trip are described in Ting Wen-chiang, "Man yu san chi," pp. 387–430. Pp. 430–437 consist of a brief history of

the Tungchuan mine. For Ting's official report on the Tungchuan mine and recommendations for reform, see *Geological Reports of Dr. V. K. Ting*, pp. 237–241.

21. Huang Chi-ch'ing, pp. 23–26.

22. Wong Wen-hao, "Tui-yü Ting Tsai-chün hsien-sheng ti chui-i" (Recollections of Mr. Ting Tsai-chün); *TLPL*, No. 188: 16 (Feb. 16, 1936).

23. Yuan Wei-chou, "Ting Tsai-chün hsien-sheng tsai ti-chih hsüeh shang chih kung-hsien" (Mr. Ting Tsai-chün's contributions to the study of geology), *CYYCY*, No. 3: 128 (December 1956).

24. V. K. Ting, "Biographical Note: A. W. Grabau," *BGSC*, 10: iii–vii (1931).

25. V. K. Ting, "Presidential Address," *BGSC*, 10.1: 9–11 (1924).

26. T'ao Meng-ho, "Chui-i Tsai-chün" (Recollections of Tsai-chün), *TLPL*, No. 188: 33–34 (Feb. 16, 1936).

27. Davidson Black, "The Geological Society and Science in China," *BGSC*, 1.1: 9–10 (March 1922).

28. Text of the speech by V. K. Ting delivered in commemoration of Davidson Black, *BGSC*, 13.3: 322 (1934).

29. Li Chi, "Tui-yü Ting Wen-chiang so t'i-ch'ang ti k'o-hsüeh yen-chiu chi tuan hui-i" (Some recollections of scientific research pioneered by Ting Wen-chiang), *CYYCY*, No. 3: 155–156 (December 1956).

30. Teilhard de Chardin, p. 138. Letter dated Feb. 20, 1927.

31. In 1936, beside the Paleontological and Cenozoic Research Laboratory, the survey was in charge of a Laboratory for the Study of Fuels, a Soil Survey, a Paleobotanical Laboratory, and a Seismological Station in Kansu. All together these organizations issued eleven publications. See W. Y. Chyne, ed., *Handbook of Cultural Institutions in China*. Shanghai, 1936, p. 185.

32. See Ting Wen-chiang, "Hsüan-hsüeh yü k'o-hsüeh," p. 24; also the chronological list of Ting's technical publications in Appendix B.

33. See in *People's China*, Li Ssu-kuang, "Science Serves the People," 4.4: 9–13 (Aug. 16, 1951); Coching Chu, "Science in New China," 1.10: 10–11 (May 16, 1950); and Kuo Mo-jo, "A New Stage in Chinese Science," No. 14: 8–12 (1955).

34. C. H. Peake, "Some Aspects of the Introduction of Modern Science into China," *Isis*, 22. 63: 173–219 (December 1934). For a balanced evaluation of Republican geology by E. C. T. Chao of the

United States Geological Survey, see Sidney H. Gould, ed., *Sciences in Communist China* (Washington, D. C., 1961), pp. 500–509.

35. Quoted by V. K. Ting, "Modern Science in China," *Asia* 36.2: 132 (February 1936).

36. Obituary of V. K. Ting, *Nature* (Jan. 18, 1936), p. 99.

37. Hu Shih, *Ting Wen-chiang ti chuan chi*, p. 33.

38. See Chap. 6, pp. 140–145.

39. Ting Wen-chiang, *Wai-tzu k'uang-yeh shih tzu-liao* (Materials for a history of foreign capitalized mining) published by the Geological Survey (Peking, 1929). Ting actually compiled the work in 1916 and 1917. See the preface by Wong Wen-hao, p. 1. Ting Wen-chiang, "Wu-shih nien lai Chung-kuo chih k'uang-yeh" (Mining in China during the past fifty years); *Tsui chin wu-shih nien* (The past fifty years). Shen pao anniversary volume (Shanghai, 1922).

40. Ting Wen-chiang, "Wu-shih nien lai Chung-kuo chih k'uang-yeh," pp. 8–9.

41. *Chung-kuo k'uang-yeh chi-yao* (Chinese mining bulletin), No. 4 (1927–1931), section on Jehol Province, pp. 272–274. However, Pei-p'iao mine was not prominent enough to be mentioned in the section on representative coal mines of the bulletin's 1926 edition.

42. Ting Wen-chiang, "Man yu san chi," pp. 371–372.

43. *Chung-kuo k'uang-yeh chi-yao*, No. 4: 272–274.

44. *Ibid.*, p. 273.

45. *Ibid.*, p. 272. See also *Far Eastern Review* (March 1926), p. 105; (May 1926), pp. 221–227.

46. Hu Shih, *Ting Wen-chiang ti chuan chi*, p. 34.

47. Ting Wen-chiang, *Wu-shih nien lai Chung-kuo chih k'uang-yeh*, pp. 3–8. See also V. K. Ting, "Mining Legislation and Development in China," *Far Eastern Review* (July 1917), pp. 570–573.

48. Ting Wen-chiang (Tsung Yen), "Ts'ai-ping chi-hua ti t'ao-lun" (Criticisms of plans for disarmament), *Nu-li chou-pao*, No. 14: 3 (Aug. 6, 1922).

4. THE NEW CULTURE AND CHINESE TRADITION

1. Fu Ssu-nien (Fu Meng-chen), "Wo so jen-shih ti Ting Tsai-chün," p. 5.

2. Ting Wen-chiang, comp., *Liang Jen-kung hsien-sheng nien-p'u*, pp. 541, 556–557. See also the preface by Ting Wen-yüan, p. 6.

3. Liang apparently took Ting's advice, for in 1922 his *Chung-kuo li-shih yen-chiu fa* (Method in the study of Chinese history) appeared.

In this work Liang tried to establish a "scientific" methodology for the study of Chinese history, by which he meant the abandonment of old classifications, refinement of techniques of textual analysis, and above all, the development of a sequential, internally logical style of discourse designed to clarify relationships of cause and effect. Though the stress on causal logic in historical explanation and on orderly classification is Tinglike, Liang maintained as well that the free, nonmaterial consciousness of man made it difficult for history to be a completely scientific subject—a view which is close to that of Chang Chün-mai, Ting's adversary in the science and metaphysics debate. See *Yin-ping-shih ho chi* (Collected works; Shanghai, 1936), Vol. 32.

4. Ting Wen-chiang, "Che-ssu-hsüeh yu p'u-tieh" (Eugenics and genealogy), *Kai-tsao* (Reconstruction), 3.4: 37–44; 5: 31–36; 6: 7–16.

5. Galton's best known book, *Hereditary Genius: An Inquiry into Its Laws and Consequences*, 2nd ed. (London, 1914), attempted to show by a statistical analysis of English families that the characteristics which produce an "eminent" individual ("ability, zeal and application") tend to be clustered in certain families and have a statistical frequency proportional to the closeness of kinship. His assumption that character traits are inherited was tested in another famous study, of identical and fraternal twins. Here Galton concluded that personality similarities were stronger among the former than among the latter. See Francis Galton, *Inquiries into Human Faculty* (London, 1883).

6. Undoubtedly Ting knew that point 7 is to be taken statistically, but his language does not make the issue unequivocally clear. Ting Wen-chiang, "Che-ssu-hsüeh yu p'u-tieh," *Kai-tsao*, 3.5: 34.

7. *Ibid.*, 3.4: 44.

8. *Ibid.*, 3.5: 36.

9. Galton himself, although apparently unaware of the scope of Chinese family records, had been enthusiastic about the possibilities of eugenics research on Chinese examination candidates. ". . . [in China] the system of examination is notoriously strict and far-reaching, and boys of promise are sure to have passed from step to step until they have reached the highest level of which they are capable . . . I feel the utmost confidence that if the question of hereditary genius were thoroughly gone into by a competent person, China would be found to afford a perfect treasury of facts bearing upon it." Galton, *Hereditary Genius*, pp. 334–335.

10. In 1922 Ting elaborated this theme in a shorter historical essay, "The Relation Between Geography and Famous Men in His-

tory," which also bears the marks of Galton's inspiration. Using data from the biographical sections of the dynastic histories, he analyzed the geographical distribution of successful examination candidates. He concluded that throughout Chinese history outstanding officials had in fact appeared in family and regional clusters, for which genealogical, as well as economic and political factors, were responsible. Leaving aside Ting's conclusions, this essay represents a pioneer attempt in the kind of statistical analysis of traditional Chinese historical materials that has since had wide applications; and it illustrates the direct way in which Ting linked biological research techniques and "scientific history." See Ting Wen-chiang, "Li-shih jen-wu yü ti-li ti kuan-hsi" (The relation between geography and famous men in history), *K'o-hsüeh*, 8.1: 10–24 (February 1922).

11. See Ju Sung, "P'ing yu-sheng-hsüeh yü huan-ching lun ti lun-cheng" (Critique of the debate on eugenics and environment), *Erh-shih shih-chi* (Twentieth century; Feb. 1, 1935).

12. Ting Wen-chih, "Wo-ti erh-ko Wen-chiang" (My second brother Wen-chiang), *TLPL*, No. 188: 49–50 (Feb. 16, 1936). See also Ting Chang Tzu-san, "Tao Tsai-chün erh-ko" (In mourning for second brother Tsai-chün), *Kuo-wen chou-pao*, 13.4: 21–22.

13. Ting Wen-chiang, "Wo-ti hsin-yang" (My beliefs), *TLPL*, No. 100: 9–12 (May 30, 1934).

14. *Ibid.*, p. 10.

15. *Ibid.*, p. 12.

16. "Ming-t'ien chiu ssu ho fang; nu-li tso ni-ti kung, chiu hsiang ni yung-yuan pu hui ssu i-yang." This pai-hua couplet, composed for Ting by Hu Shih, probably derives from the Western aphorism attributed to Isadore of Seville: "Study as if you were to live forever; live as if you were to die tomorrow." In Chinese hands the Christian maxim has undergone a significant change in emphasis. See *Wen-hsing*, No. 63: 55, a review by Li Ao of Hu's biography of Ting.

17. For details on Ting's personal outlook, see Fu Ssu-nien (Fu Meng-chen), "Wo so jen-shih ti Ting Wen-chiang."

18. Such premises lay behind Marxist versions of history as well—which were, after all, drawn from Marx's own social interpretations of Darwinian biology—and so permeated the contemporaneous debates on Chinese social history of a contrasting group of "scientific" historians following Kuo Mo-jo, Li Chi, and T'ao Hsi-sheng. See D. W. T. Kwok, *Scientism in Chinese Thought, 1900–1950* (New Haven, 1965), pp. 162–170.

19. Ting Wen-chiang, comp., *Liang Jen-kung hsien-sheng nien-p'u*, pp. 10–11.

20. Ting was particularly indignant when he found what he thought to be traditionalist assumptions cropping up in a recent Western work like Granet's *La civilization chinoise*. By treating a vast and contradictory corpus of ancient texts as if they were undifferentiated, and by constructing from them a portrait of a quintessential Chinese civilization of the Chou dynasty, Granet seemed guilty of errors from which Chinese scholars were just learning to emancipate themselves. See V. K. Ting, "Professor Granet's 'La civilization chinoise,'" *Chinese Social and Political Science Review* (July 1931).

21. V. K. Ting, "How China Acquired her Civilization," in *Symposium on Chinese Culture*, ed. Sophia H. Chen Zen (Shanghai, 1931), p. 26.

22. Liang Ch'i-ch'ao, *Intellectual Trends in the Ch'ing Period*, tr. Immanuel C. Y. Hsü (Cambridge, Mass., 1959).

23. Hu Shih, *The Development of the Logical Method in Ancient China* (Shanghai, 1922).

24. Hu Shih, *The Chinese Renaissance* (Chicago, 1934), pp. 63–77. See also Liang Ch'i-ch'ao, pp. 14–15. This entire discussion, of course, owes much to the insights of Joseph Levenson, *Confucian China and Its Modern Fate*, Vol. 1.

25. V. K. Ting, "On Hsu Hsia-k'o (1586–1641), Explorer and Geographer," pp. 326–327. This article was first read as a paper before the "Friends of Literature Society" (*wen yu hui*) in Peking on June 14, 1921.

26. Hu Shih, "Ch'ing-tai hsüeh-che ti che-hsüeh fang-fa" (The philosophical method of Ch'ing scholars), in *Hu Shih wen-tsun* (Collected works), Ser. 3, II, 539–580.

27. Ting Wen-chiang, "Hsü Hsia-k'o yu chi," *CYYCY*, No. 3: 513–522. This essay originally appeared in the *Hsiao-shuo yüeh-pao* (Short story monthly), Vol. 17, special supplement on Chinese culture (1926). It also served as the preface to Ting's edition of Hsü's travel diaries published in 1929.

28. *Ibid.*, pp. 515–516.

29. V. K. Ting, "On Hsu Hsia-k'o," p. 337.

30. Sung Ying-hsing, *T'ien kung k'ai wu* (The creations of nature and human labor), ed. T'ao Hsiang, with a critique and biography of the author by Ting Wen-chiang (Shanghai, 1929). Ting's preface and appendix are reprinted in *CYYCY*, No. 3: 523–528. A translation of the

newly rediscovered Ming edition has been published in English by E-tu Zen Sun and Shiou-chuan Sun as *T'ien kung k'ai wu, Chinese Technology in the Seventeenth Century* (University Park and London, 1966).

31. Ting Wen-chiang, "Feng-hsin Sung Chang-keng chuan," (Biography of Mr. Sung Chang-keng of Feng-hsin), *CYYCY*, No. 3: 523.

32. V. K. Ting, "How China Acquired her Civilization," pp. 24–25.

33. An interesting evaluation of the methodology of Ch'ing classical scholars may be found in Bernard Karlgren, "The Authenticity of Ancient Chinese Texts," *Bulletin of the Museum of Far Eastern Antiquities*, No. 1: 165–183 (1929). Karlgren gives numerous instances of inductive reasoning by Chinese textual scholars, and he cites as typical faults drawing conclusions on the basis of incomplete evidence and lapses into the "higher criticism."

34. Both Ting and Hu Shih were aware that there was a philosophical "problem of induction" which complicated contemporary discussions of scientific method in the West: since the researcher can never examine all possible cases before making a generalization with the presumed status of scientific law, he cannot be certain that his generalization is truly lawlike and universal. Both men made passing references to European philosophers of science who tried to deal with this difficulty. Hu turned to revisers of J. S. Mill like W. Stanley Jevons, who posited a combined "inductive-deductive" method for science; Ting turned to positivist axioms concerning the "uniformity of nature" and later to early probability theory. Such theories were in fact modifications of the idea of an inductive logic of science, but not repudiations of its base. See Hu Shih, "Ch'ing-tai hsüeh-che ti fang-fa," in *Hu Shih wen-tsun*, Ser. 3, II, 539–541; and Ting Wen-chiang, "Hsüan-hsüeh yü k'o-hsüeh," p. 10.

35. For details, see Chap. 5.

5. SCIENCE AND METAPHYSICS

1. Wu Chih-hui, "Chen yang pa-ku-hua chih li-hsüeh" (A warning against neo-Confucian philosophy in Western pa-ku disguises), in *KHY*.

2. For Ting's political activities between 1921 and 1923, see Chap. 6.

3. Liang Ch'i-ch'ao, "O yu hsin-ying lu chieh lu" (Reflections on a trip to Europe), in *Yin-ping-shih ho chi*, Vol. 21.

4. Liang Sou-ming, *Tung hsi wen-hua chi ch'i che-hsüeh* (Eastern and Western culture and their philosophies; Shanghai, 1921).

5. Chang Chün-mai, "Jen-sheng kuan" (A view of life), in *KHY*. Short selections from Chang's essay, as well as from Ting's reply and

other contributions to the debate, have been translated in W. Theodore de Bary, ed., *Sources of Chinese Tradition* (New York, 1960), pp. 834–843.

6. Bergson argued that science considers time only as a dimension like a spatial dimension, whose essential properties can be depicted geometrically. It would be more accurate, he suggested, to see time as a dimension of consciousness, which perceives it as duration. Duration, then, would be an essential aspect of human experience, but since it cannot be divided into fixed quantities, it is immune to scientific analysis. In his capacity as "chose qui dure," man enjoys free will and is liberated from the straitjacket of scientific law. For Chang's handling of this thesis, see his "Tsai lun jen-sheng kuan yü k'o-hsüeh ping ta Ting Tsai-chün" (More discussion of a view of life and science, with a reply to Ting Tsai-chün), in *KHY*, p. 65.

7. Hans Driesch (1867–1941), considered both a philosopher and a biologist, attempted to quarantine the spread of mechanistic scientific methods by postulating a quasi-mystical, quasi-organic "entelechy," "life" itself, as a property of all biological organisms. He offered his own experiments in embryology as proof of his thesis. Carsun Chang accompanied Driesch on a lecture tour through China in 1923, and Chang's original speech on "A View of Life" actually was delivered in honor of Driesch's visit. See Carsun Chang, "Reflections on the Philosophical Controversy in 1923," *Chung Chi Journal*, 3.1: 19 (November 1963).

8. Chang's division of the sciences into the categories of "physical" and "spiritual" was adopted from the "Exaktewissenschaft" and "Geisteswissenschaft" of the German philosopher Wundt. Wundt in turn had proposed them as modifications of traditional German academic classifications. Chang did not follow Wundt exactly, for where Wundt held biology to be an exact science, Chang's vitalist beliefs left him uncertain as to its proper status. See Chang Chün-mai, "Tsai lun jen-sheng kuan yü k'o-hsüeh," in *KHY*, pp. 4–31.

9. In Chang's words, "Material substance in space is easily experimented upon, while 'vital forces' [English used] of biology are correspondingly difficult to experiment upon, and psychology more difficult still . . . It is easy to determine the temporal sequence of material events and know cause and effect. In the world of biology temporal sequence may be distinguishable, yet it is not easy to know the cause of the formation of the embryo; since in psychology the mind is continually changing, to an even greater extent there are no fixed conditions to be sought . . . Because both biology and psychology are subject to the difficulty of

constant differences between individuals, it is difficult to achieve 'uniformity' [English used]. Chang Chün-mai, "Tsai lun jen-sheng kuan yü k'o-hsüeh," in *KHY*, pp. 23–24.

10. *Ibid.*, pp. 77–79.

11. Carsun Chang, "Reflections on the Philosophical Controversy in 1923," p. 21.

12. Cf. Wang Yang-ming, "*Liang chih* is the compass, the square and the measure. On the basis of the subtle stirrings of *liang chih*, one can discover whether one is headed towards the right or the wrong, and it is in paying heed to these small warnings that one should be most strict with oneself." The translation is Chang's own, from his volume in English, *Wang Yang-ming* (Jamaica, N. Y., 1962), p. 46. Chang thought that he had found in the idealist spiritualism of Rudolf Eucken a European parallel to the doctrines of Sung-Ming metaphysics. He saw in Eucken's dichotomy between spirit and matter correspondences with the neo-Confucian categories of "inner" and "outer" (*nei* and *wai*): "Man's life in the world is called spiritual respecting that which is is within, and material respecting that which is without. That which is spiritual within is in motion and does not rest; that which is material without is static and does not advance." See Chang Chün-mai, "Tsai lun jen-sheng kuan yü k'o-hsüeh," in *KHY*, p. 31.

13. The most eloquent statement of this came not from Chang himself, but from Liang Ch'i-ch'ao in "O yu hsin-ying lu chieh lu": "[Europeans] take their inner life and make it correspond with their outer. Large numbers of strangers are gathered together to live in a city or factory. Outside of concern for material benefits they have no feelings for each other to speak of . . . After work they seek enjoyment, but before they have finished enjoying themselves they must work again; they are busy day and night with no resources left for self improvement . . . Their desires daily increase, things daily grow more expensive, life daily grows more difficult and competition more fierce." (p. 10)

14. Ting Wen-chiang, "Hsüan-hsüeh yü k'o-hsüeh," in *KHY*, p. 2.

15. *Ibid.*, p. 18. The quotation marks indicate Ting's quotations from Chang's original article.

16. Karl Pearson, *The Grammar of Science* (Everyman's Library, London, 1937), first published in 1892. Pearson (1857–1936) was a professor of applied mathematics and later of eugenics at the University of London and the chief follower in England of Francis Galton. His own most important scientific work lay in the field of biological and social statistics. The physicist Ernst Mach was the other chief exponent of this

kind of epistemology of science, which was often referred to in contemporary writings as "sensationalism." See Ernst Mach, *The Analysis of Sensations*, translated from the first German edition by C. M. Williams; revised and supplemented from the fifth German edition by Sidney Waterlow (Chicago and London, 1914). Although Ting often referred to Mach, in his writings for Chinese audiences he relied far more heavily on the arguments of Pearson. Moreover, Ting's views on genetics and scientific sociology were remarkably close to those of Pearson, who believed that eugenic studies should provide the foundation for social reforms designed to further the biological improvement of the species. See especially Pearson's *Nature and Nurture: the Problem of the Future* (London: University of London, Galton Laboratory for National Eugenics, 1910).

17. Ting Wen-chiang, "Hsüan-hsüeh yü k'o-hsüeh," in *KHY*, pp. 8–9.

18. *Ibid.*, pp. 9–10.

19. *Ibid.*, p. 13. Pearson wrote, "We are like the clerk in the central telephone exchange who cannot get nearer to his customers than his end of the telephone wires. We are indeed worse off than the clerk, for to carry out the analogy properly we must suppose him never to have seen a customer—in short, never, except through the telephone wire, to have come in contact with the outside universe." *Grammar of Science*, p. 56.

20. Ting Wen-chiang, "Hsüan-hsüeh yü k'o-hsüeh," in *KHY*, pp. 12–13.

21. *Ibid.*, pp. 9–10.

22. Thomas Huxley, "On Descartes' 'Discourse Touching the Method of Using One's Reason Rightly and of Seeking Scientific Truth,'" *Lay Sermons, Addresses and Reviews* (London, 1899). On this point Huxley wrote, "The reconciliation of physics and metaphysics lies in the acknowledgement of faults on both sides; in the confession by physics that all the phenomena of nature are, in their ultimate analysis, known to us only as facts of consciousness; in the admission by metaphysics that facts of consciousness are, practically, interpretable only by the methods and formula of physics." pp. 299–300.

23. Pearson stated his position as follows: "Beyond the sense impression, beyond the brain terminals of the sensory nerves, we cannot get. Of what is beyond them, of 'things-in-themselves' as the metaphysicians term them, we can know but one characteristic, and this we can only describe as a capacity for producing sense impressions, for sending messages along the sensory nerves to the brain. This is the sole scientific statement which can be made with regard to what lies beyond sense

impressions." *Grammar of Science*, p. 61.

24. Benjamin Schwartz, *In Search of Wealth and Power: Yen Fu and the West* (Cambridge, Mass., 1964). See pp. 186–196.

25. That Ting himself had not completely understood the Cartesian implications of "skeptical idealism" may be seen from an exchange with Lin Tsai-p'ing, a Buddhist. Lin, interestingly, was the only one who questioned the idea pertinently: "If we say we know things by means of sense perceptions and we call this scientific knowledge, then things we perceive in a dream cannot be said to have a different value from things we perceive when awake. Can it be that scientific knowledge is like our state of mind during dreams?" But Ting refused to answer this point, saying that the Buddhist and scientific philosophies were too fundamentally opposed to be debated profitably. See Lin Tsai-p'ing, "Tu Ting Tsai-chün hsien-sheng ti 'hsüan-hsüeh yü k'o-hsüeh'" in *KHY*, p. 19; and Ting Wen-chiang, "Hsüan-hsüeh yü k'o-hsüeh ti t'ao-lun ti yü-hsiang," in *KHY*, p. 4.

26. Ting Wen-chiang, "Hsüan-hsüeh yü k'o-hsüeh," in *KHY*, p. 20. Cf. Pearson: "The field of science is unlimited; its material is endless; every group of natural phenomena, every phase of social life, every stage of past and present development, is material for science. The unity of science lies alone in its method, not in its material." *Grammar of Science*, p. 16.

27. *Ibid.*, p. 99. For Pearson's account of scientific law and causation, see Chaps. III and IV, pp. 68–130.

28. Ting Wen-chiang, "Hsüan-hsüeh yü k'o-hsüeh: Ta Chang Chün-mai," *KHY*.

29. Ting Wen-chiang, "Hsüan-hsüeh yü k'o-hsüeh," in *KHY*, p. 3. On this point Pearson wrote, "The scientific method . . . consists in the careful and often laborious classification of facts, in the comparison of their relationships and sequences, and finally in the discovery by aid of the disciplined imagination of a brief statement or formula which in a few words resumes a wide range of facts. Such a formula . . . is termed a scientific law." *Grammar of Science*, p. 69.

30. Thomas Huxley, "The Method of Zadig," *Science and the Hebrew Tradition* (London, 1910). See also Hu Shih, *Ting Wen-chiang ti chuan chi*, p. 50.

31. Ting Wen-chiang, "Hsüan-hsüeh yü k'o-hsüeh," in *KHY*, p. 20; "Hsüan-hsüeh yü k'o-hsüeh: ta Chang Chün-mai," in *KHY*, p. 16.

32. See Stephen Toulmin, *The Philosophy of Science: an Introduction* (London & New York, 1953). In explaining the nature of a theoretical

concept in science Toulmin uses the example of a "light ray" in geometrical optics. The basic theory states that "light travels in straight lines," and it is better stated by the geometrical model of diagrams showing the behavior of light rays than by any linguistic approximations. The term "light ray" is a language substitute for the geometrical ideal of a straight line, and it in no way describes a sunbeam, or even a beam of light cast by an experimental optical apparatus, both of which will be sharp only to a certain point. Moreover, scientific theories viewing light in different aspects, such as those which deal with the wave theory of how light travels, will abandon the concept of a "light ray" entirely. Under these circumstances "light ray" has a meaning and use quite removed from any phenomenon experienced in ordinary life. See especially Chap. II, pp. 17–56.

33. Ting Wen-chiang, "Hsüan-hsüeh yü k'o-hsüeh," in *KHY*, pp. 28–29.

34. Wu Chih-hui, "Chen yang pa-ku-hua chih li-hsüeh," in *KHY*, p. 4.

35. Wu Chih-hui, "I-ko hsin hsin-yang ti yü-chou kuan yü jen-sheng kuan" (A new belief and its view of life and of the universe), in *KHY*, pp. 22–23.

36. *Ibid.*, p. 35.

37. *Ibid.*, pp. 9–10.

38. *Ibid.*, p. 1.

39. Hu Shih, *The Chinese Renaissance*, pp. 91–92.

40. Hu Shih, "K'o-hsüeh yü jen-sheng kuan hsü" (Preface to Science and a View of Life), in *KHY*, pp. 25–27. Paraphrase of the Chinese. For an English version of these remarks, see Hu Shih, "What I Believe," *Forum Magazine*, February 1931.

41. *Ibid.*, pp. 27–29.

42. Ch'en Tu-hsiu, "K'o-hsüeh yü jen-sheng kuan hsü" (Preface to Science and a View of Life), in *KHY*, p. 11.

43. Ting Wen-chiang, "Hsüan-hsüeh yü k'o-hsüeh," in *KHY*, pp. 20–21.

44. Ting Wen-chiang, "Hsüan-hsüeh yü k'o-hsüeh: Ta Chang Chün-mai," in *KHY*, pp. 39–42.

45. Ting Wen-chiang, "Hsüan-hsüeh yü k'o-hsüeh," in *KHY*, pp. 21–26.

46. For example, the expression "jen-sheng kuan" itself was a fertile source of complications, for it was used by participants in a number of ways which, though related, should not be identified with one

another. An important ambiguity is rendered by the shift in emphasis between the alternative English expressions "outlook on life," implying an individual's subjective feelings, and "view of man's life," implying an estimate of the human condition in totality. When Ting said, "A jen-sheng kuan is a man's knowledge and emotions, plus his attitude toward knowledge and emotions," he was using the expression in a psychological sense, referring to an individual's point of view. ("Ta Chang Chün-mai," in *KHY*, p. 41). Hu Shih and Wu Chih-hui, when they aspired to a "scientific jen-sheng kuan," were presenting the term in the second sense of an objective appraisal; and when Fan Shou-kang wrote that "in the main a jen-sheng kuan is made up of the laws of ethics" ("Ping so-wei k'o-hsüeh yü hsüan-hsüeh chih cheng," in *KHY* (p. 14), he was implying that it was an ideal, a model for human conduct. Chang adopted the expression from Rudolf Eucken's "Lebensanschauung," which is close to "philosophy of life," a term that usually has a subjective connotation in the West.

47. Chang Chün-mai, for example, later said that the most representative participants in the debate had been Hu Shih, Wu Chih-hui, and Ch'en Tu-hsiu. See Chang Chün-mai, "Jen-sheng kuan lun-chan chih hui-ku" (Looking back on the debate over a view of life), *Tung-fang tsa-chih*, 31, 13: 5–13 (July 1, 1934).

48. Ting Wen-chiang, "Hsüan-hsüeh yü k'o-hsüeh ti t'ao-lun ti yü-hsing," in *KHY*. Ting's final remarks on the subject took the wry form of a book list for further study. The list was composed of books largely from his own library. In addition to works on biology, physics, chemistry, and anthropology, he recommended the following titles on general issues of science and philosophy: Sedgwick and Tyler, *A Short History of Science;* J. T. Merz, *History of European Thought in the Nineteenth Century;* Karl Pearson, *The Grammar of Science;* W. S. Jevons, *The Principles of Science;* T. Huxley, *Method and Results*, and *Science and Education;* Veblen, *The Place of Science in Modern Civilization;* F. Soay, *Science and Life;* Robinson, *The Mind in the Making;* W. James, *The Principles of Psychology* and *Textbook in Psychology;* Notsworthy and Whitely, *The Psychology of Childhood;* S. J. Holmes, *The Evolution of Animal Intelligence;* E. Mach, *The Analysis of Sensations;* B. Russell, *The Analysis of Mind;* J. Dewey, *Reconstruction in Philosophy*, *Essays in Experimental Logic*, and *German Philosophy and Politics;* H. Bergson, *Creative Evolution* and *Mind and Energy* [L'Energie spirituelle]; H. M. Kallen, *William James and Henri Bergson;* M. S. R. Elliot, *Modern Science and the Illusions of Professor Bergson.*

49. Mousheng Hsitien Lin, "Recent Intellectual Movements in

China," *China Institute Bulletin*, 3.1: 9 (October 1938).
  50. Teilhard de Chardin, pp. 108–109.

6. *ENDEAVOR:* A POLITICAL EDUCATION

  1. Hu Shih, et al. "Cheng tzu-yu ti hsüan-yen" (Manifesto of struggle for freedom), *Tung-fang tsa-chih* 17.16: 133–134 (Aug. 25, 1920).

  2. "Wo-men ti cheng-ch'ih chu-chang" (Our political proposals), *Nu-li chou-pao*, No. 2 (May 14, 1922). See *ibid.*, No. 3, for a list of the additional signatories, who were: Li Chien-tung (Principal of Peking Upper Normal School), Mao Pang-wei (Principal of Peking Women's Upper Normal School), Wang Chia-chü and Yü T'ung-k'uei (both listed as Principal of Peking Law School), Chou Sung-sheng (Principal of Peking Medical School), Wu Tsung-chih (Principal of Peking Agricultural College), and Yeh Chin (Principal of Peking Art School).

  3. Hu Shih, *Ting Wen-chiang ti chuan chi*, pp. 35–36, 40.

  4. Ting Wen-chiang (Tsung Yen), "I-ko wai-kuo p'eng-yu tui-yü i-ko liu-hsüeh-sheng ti chung-kao" (Counsels of a foreign friend to a Chinese graduate from abroad), *Nu-li chou-pao*, No. 42 (Mar. 4, 1923).

  5. Hu Shih, *Ting Wen-chiang ti chuan chi*, p. 35.

  6. "Wo-men ti cheng-ch'ih chu-chang," *Nu-li chou-pao*, No. 2.

  7. *Ibid.*

  8. *Ibid.*

  9. Hu Shih, "Kuan-yü 'wo-men ti cheng-ch'ih chu-chang' ti t'ao-lun" (On criticisms of "our political proposals"), *Nu-li chou-pao*, No. 4 (May 28, 1922). Hu Shih's reply to Comment 9.

  10. *Ibid.*

  11. Ting Wen-chiang (Tsung Yen), "Ta kuan-yü 'wo-men ti cheng-ch'ih chu-chang'" (Reply concerning "our political proposals"), Pt. 1, *Nu-li chou-pao*, No. 6: 3–4 (June 11, 1922).

  12. Ting Wen-chiang (Tsung Yen), "Ta kuan-yü 'wo-men ti cheng-ch'ih chu-chang,'" Pt. 2, *Nu-li chou-pao*, No. 7 (June 18, 1922), p. 3.

  13. *Ibid.*

  14. Hu Shih, "Kuan-yü 'wo-men ti cheng-ch'ih chu-chang' ti t'ao-lun," *Nu-li chou-pao*, No. 4.

  15. Ting Wen-chiang (Tsung Yen), "Ta kuan yü 'wo-men ti cheng-ch'ih chu chang,'" p. 3.

  16. Hu Shih, "Kuan-yü 'wo-men ti cheng-ch'ih chu-chang' ti t'ao-lun."

  17. Ting Wen-chiang (Tsung Yen), "Ta kuan yü 'wo-men ti cheng-

ch'ih chu chang,'" *Nu-li chou-pao*, No. 6.

18. Ting Wen-chiang (Tsung Yen), "Chung-kao chiu kuo-hui i-yüan" (Counsels to the deputies of the old parliament), *Nu-li chou-pao*, No. 9 (July 2, 1922).

19. Li Chien-nung, *The Political History of China: 1840–1928*, tr. Ssu-yu Teng and Jeremy Ingalls (Princeton, 1956), pp. 419–435.

20. *Nu-li chou-pao*, No. 30 (Nov. 26, 1922), p. 2, quoting the *Ch'en pao* (Nov. 22, 1922).

21. Editorial signed by Hu Shih and Ting Wen-chiang (Tsung Yen), *Nu-li chou-pao*, No. 30. See also Kao I-han, "Kuo fa ho tsai!" (Where is the law of the land!), *ibid.*, p. 2.

22. *Nu-li chou-pao*, No. 39 (Jan. 28, 1923). Ts'ai Yüan-p'ei's statement of resignation.

23. Ts'ai Yüan-p'ei, "Wo tsai Peiching ta-hsüeh ti ching-li" (My presidency of Peking university), *Tung-fang tsa-chih*, Vol. 31, No. 1 (Feb. 14, 1934).

24. *Nu-li chou-pao*, No. 38 (Jan. 21, 1923). Hu Shih's article on Ts'ai Yüan-p'ei's resignation.

25. Quoted by Hu Shih in *Ting Wen-chiang ti chuan chi*, p. 40.

26. Hu Shih, "Shih-yen chu-i" (Experimentalism), in *Hu Shih wen-ts'un*, Vol. 2, pp. 409–479.

27. For this insight I am indebted to Professor Lin Yu-sheng, who analyzes continuities in the categories of analysis of Chinese intellectuals in an unpublished essay, "The Cultural Intellectualistic Approach of the Chinese Intelligentsia."

28. See especially Hu Shih, "Kuan-yü 'wo-men ti cheng-ch'ih chu-chang' ti t'ao-lun," *Nu-li chou-pao*, No. 4 (May 28, 1922).

29. Ting Wen-chiang (Tsung Yen), "Shao-shu jen ti tse-jen" (The responsibility of a few men), *Nu-li chou-pao*, No. 67 (Aug. 26, 1923).

30. *Ibid.*

31. *Ibid.*

7. DIRECTOR-GENERAL OF GREATER SHANGHAI

1. See *Nu-li chou-pao*, Nos. 1, 3, 5, 19, and 28.

2. Ting Wen-chiang, *Min-kuo chün-shih chin-chi* (Notes on republican military affairs; Shanghai, 1926).

3. Hu Shih, *Ting Wen-chiang ti chuan chi*, p. 38.

4. Ting Wen-chiang (Tsung Yen), "Ts'ai-ping chi-hua ti t'ao lun," *Nu-li chou-pao*, No. 14: 2–4. Chiang's book was entitled *Ts'ai-ping chi-hua shu* (Plan for disarmament). Ting found serious fault with its

provisions which would in effect retain the existing officer's corps on pension while reducing the ranks and with its suggestions for the resettlement of veterans, which he thought seriously underestimated the economic problems involved. Since these objections undercut the substance of Chiang's proposals, Ting's kindness as a reviewer probably is at least in part a reflection of his friendship with the author.

5. Quoted by Tsiang T'ing-fu, "Wo so chi-te ti Ting Tsai-chün" (The Ting Tsai-chün I remember), *CYYCY*, No. 3: 139–140.

6. Hu Shih, *Ting Wen-chiang ti chuan chi*, pp. 61–62.

7. Tsiang T'ing-fu, "Wo so chi-te ti Ting Tsai-chün," p. 146.

8. Ting Wen-chiang, *Min-kuo chün-shih chin-chi*, pp. 147–149. See also Ting Wen-chiang, *Man yu san chi*, p. 365.

9. Donald G. Gillin, *Warlord: Yen Hsi-shan in Shansi Province, 1911–1949* (Princeton, 1967).

10. Ting Wen-chiang, *Min-kuo chün-shih chin-chi*, pp. 182–183.

11. Hu Shih, *Ting Wen-chiang ti chuan chi*, p. 34.

12. *Ibid.*, p. 41. See also Fu Ssu-nien (Fu Meng-chen), "Ting Wen-chiang i-ko jen-wu," p. 9.

13. Since at that time Lo was in the service of the Peking government, and since this was not the only occasion when he communicated secretly with Ting, it seems likely that, under cover of his position, he was supplying intelligence to Chang Tso-lin's adversaries. See below, p. 187.

14. Hu Shih, *Ting Wen-chiang ti chuan chi*, p. 41.

15. Tung Hsien-kuang, p. 136.

16. *The China Yearbook, 1926–1927*, pp. 928–929.

17. Since the Washington Conference had declared that in the interpretation of agreements granting rights and privileges to foreigners, "such grants shall be strictly construed in favor of the grantors," its directives had radical implication for an authority like the Shanghai Municipal Council, which had long governed by means of precedents based on indistinctly worded texts interpreted invariably in favor of the grantee. See A. M. Kotenev, *Shanghai: Its Municipality and the Chinese* (Shanghai, 1927). See esp. p. 26.

18. *The China Yearbook, 1926–1927*. See pp. 931–933 for the English text of the letter. Signers were Liang Ch'i-ch'ao, Chu Chi-chien, Wellington Koo, Chang Kuo-kan, Fan Yuan-lien, Li Shih-wei, Hollington Tong, and Ting.

19. Lo Chia-lun, "Hsien-tai hsüeh-jen Ting Tsai-chün ti i-chüeh" (An essay on the contemporary scholar Ting Tsai-chün), *CYYCY*, No. 3: 151.

20. *The China Yearbook, 1926–1927*, p. 941.

21. An incident which occurred in connection with the restoration of the British concessions at Hankow and Kiukiang in February 1927 strikingly illustrates the way in which Ting underestimated the effectiveness of brute power in imperialist conflicts. If we are to believe the testimony of a British friend, after Chinese mobs had stormed the Hankow concession, Ting "went to Nanking [i.e., to Sun Ch'uan-fang] to propose to his Government a scheme which would satisfy both British and Chinese demands, and was distressed when, after apparently reaching a satisfactory solution, the announcement was made in the House of Commons that Great Britain had given up Hankow." See the obituary of V. K. Ting, *Nature* (Jan. 18, 1936), p. 99.

22. Great Britain had resolved in 1922 to channel its Boxer payments into a beneficial project in China; the commission, charged with making recommendations concerning the disposal of this money, was organized in 1925. Its chairman was Earl Buxton, the vice-chairman was Viscount Willingdon; English members were Dame Adelaide Anderson, W. E. Soothill, Sir C. Needham, and Sir Charles Addis; the Chinese were represented by Ting, Hu Shih, and Wang Ching-ch'un, a director of the Chinese Eastern Railway. A subcommittee headed by Willingdon and including the three Chinese conducted an on-the-spot investigation in China beginning in February 1926, in Shanghai. Ting participated in the subcommittee's work only until his post in Shanghai intervened, and he did not accompany the others to London that summer to draw up a final report. The report asked for the establishment of a permanently endowed, independent, Chinese philanthropic foundation. Ting's hand may be seen most prominently in the recommendation that 26 percent of the foundation's income be allocated to "scientific research"—the largest share to go to a National Research Institute for postgraduate study in geology, paleontology, zoology, botany, prehistory, history, anthropology, physics, and chemistry. Although this proposal was first made to the subcommittee by Ts'ai Yüan-p'ei, Ting drew up a prospective budget for the institute which appeared in the final report, and his colleague Wong Wen-hao of the Geological Survey contributed a special memorandum on it. *Report of the Advisory Committee Together With Other Documents Respecting the China Indemnity. British Parliamentary Papers, China No. 2, 1926.* See also Hu Shih, *Ting Wen-chiang ti chuan chi*, pp. 59–60; and Wang Ching-chun, *Japan's Continental Adventure* (New York, 1941), pp. 130–131.

23. See Hu Shih, *Ting Wen-chiang ti chuan chi*, p. 61. Hu Shih himself,

while defending Ting's patriotic motives, offers few details concerning his friend's decision—a reticence which is especially noticeable considering the fact that Hu was in Shanghai with Ting at the time and even attended the banquet which inaugurated the new directorate. See *Shen pao* (May 6, 1926).

24. A. M. Kotenev, *Shanghai: Its Mixed Court and Council* (Shanghai, 1925), p. 274. A particularly well publicized example of court abuse occurred during Ting's administration and was embarrassing for his authority, although stimulating to the public campaign for rendition. During a visit to Shanghai in June 1926, Hsiung Hsi-ling, a former prime minister of the republic, was hauled into court by the American consul. The plaintiff, an American engineer, wanted to hold Hsiung responsible for salary arrears owed him by a bankrupt Chinese mining company in Honan, of which Hsiung had formerly been a director. Even Ting's personal intervention failed to secure a dismissal of the charge, which left the Chinese dignitary no alternative but to jump bail. See *Shen pao* (June 29, 1926); and *Hsiang tao* (July 14, 1926).

25. *Shen pao* (July 17 and Sept. 28, 1926). For English text of the new agreement and regulations, see Kotenev, *Shanghai: Its Municipality and the Chinese*, pp. 186–191.

26. V. K. Ting, "The Greater Shanghai Municipality," *North China Herald* (Oct. 16, 1926).

27. Sun Ch'uan-fang's speech before the Chinese Chamber of Commerce on May 5, 1926. See *Shen pao* (May 6, 1926). For English text, see *North China Herald* (May 8, 1926).

28. *Shen pao* (May 16, May 17, June 24, July 20, Aug. 7, Nov. 6, 1926). *North China Herald* (May 22, Aug. 21, Aug. 28, Sept. 4, Oct. 16, and Oct. 30, 1926).

29. Ting, "The Greater Shanghai Municipality."

30. *North China Herald* (May 29, June 5, 1926). *Hsiang tao* (June 3, 1926).

31. For the directorate's position on the strike movement, see *Shen pao* (June 30 and July 16, 1926); for the rice crisis, see *ibid.* (June 22, June 23, July 10, Aug. 24).

32. The available record does reveal one instance when Ting sought to mitigate the political repressions of the Shanghai authorities. In August 1926 some writers associated with the Creation Society were arrested by police who wanted to close down their press on grounds of revolutionary subversion. Ting intervened and arranged their release from jail. See *Hung shui*, 2.23–24: 577–580.

33. *Kuo-wen chou-pao* (July 18, 1926). See also Fu Ssu-nien (Fu Meng-chen), "Ting Wen-chiang i-ko jen-wu," p. 10.

34. F. F. Liu, *A Military History of Modern China* (Princeton, 1956), pp. 36–39.

35. *Shen pao* (Sept. 11, Oct. 3, 1926); *North China Herald* (Sept. 4, 1926).

36. *Shen pao* (Sept. 15, Sept. 16, 1926).

37. *Ibid.* (Oct. 17–25, 1926). It is said that Sun Ch'uan-fang received advance warning of this attack from Ch'en I, who was rewarded with the Chekiang governorship after Hsia's defeat.

38. *Ibid.* (Nov. 20, Nov. 23, 1926).

39. *Ibid.* (Nov. 29, Dec. 4, 1926); *North China Herald* (Dec. 4, Dec. 11, Dec. 18, 1926).

40. Hu Shih, *Ting Wen-chiang ti chuan chi*, pp. 66–68.

41. Fu Ssu-nien (Fu Meng-chen), "Ting Wen-chiang i-ko jen-wu," p. 10.

42. *Shen pao* (Dec. 12–13, 1926). For Ch'en T'ao-i's retirement on pretext of personal business, see *ibid.* (Dec. 6, 1926).

43. Fu Ssu-nien (Fu Meng-chen), "Ting Wen-chiang i-ko jen-wu," p. 11.

44. *Ibid.*, p. 9.

45. Cf. Hu Shih in *Ting Wen-chiang ti chuan chi*, p. 66. See also Tung Hsien-kuang, pp. 136–137.

46. Fu Ssu-nien (Fu Meng-chen), "Ting Wen-chiang i-ko jen-wu," p. 9.

47. *Hsiang tao* (July 1, July 14, Oct. 4, 1926). See especially the issue of June 30, for an article by Ch'en Tu-hsiu on the reactionary influence of Westernization in Shanghai, in which he concluded, "Ting Wen-chiang and Sun Ch'uan-fang want a 'Greater Shanghai'; we want a 'Revolutionary Shanghai,' because a Shanghai under foreign slavery will be more rotten the greater it is." pp. 1568–1569.

8. THE INDEPENDENT CRITIC

1. A fortunate gift eased Ting's financial circumstances. His bene-factor was Yang Chü-ch'eng, the wealthy proprietor of several flour mills, whom Ting had first known as a destitute peasant in Peking. As a child, Mr. Yang had been taken to the United States by an American mission-ary, and before returning home he acquired valuable experience working in American mine fields. Ting found him a skilled job with a Chinese mining company and in this way started Yang on the road to fortune.

The gift, of five thousand yuan, ostensibly came from Mr. Yang alone, but some of Ting's students and colleagues probably contributed also. See Hu Chen-hsing, "Shui sung chi Ting Wen-chiang wu ch'ien yüan?" (Who gave Ting Wen-chiang five thousand yüan?), *TLPL*, No. 193:8–9.

2. Li Chi, pp. 156–157.

3. On this trip he did most of the research for "The Orogenic Movements in China," *BGSC*, 8.2:151–170 (1928); and for "The Permian of China and Its Bearing on Permian Classification," with A. W. Grabau, Report 15, International Geological Congress, Washington, D. C., 1933. The technical field notes Ting amassed on this trip may be found in *Ting Wen-chiang hsien-sheng ti-chih t'iao-ch'a pao-kao*, pp. 275–370.

4. Quoted in Hu Shih, *Ting Wen-chiang ti chuan chi*, p. 73.

5. Ling Hung-hsün, "Hui Ting Wen-chiang hsien-sheng ping chi ch'i tui-yü t'ieh-lu ti i-chien" (Recollections of Ting Wen-chiang and of his views on railroads), in Hu Shih et al. *Ting Wen-chiang che-ko jen* (This man, Ting Wen-chiang; Taipei, 1968), pp. 217–224.

6. Hu Shih, *Ting Wen-chiang ti chuan chi*, pp. 74–75. See also Ting Chang Tzu-san, "Tao Tsai-chün erh-ko," p. 21.

7. Chiang Monlin (Chiang Meng-lin), *Tides from the West; A Chinese Autobiography* (New Haven, 1947), p. 202.

8. Yüan Wei-chou, p. 127. See also Kao Chen-hsi, "Tso chiao-shih ti Ting Wen-chiang hsien-sheng" (Mr. Ting Wen-chiang the teacher), *TLPL*, No. 188:50–52.

9. The maps were drawn on a scale of 2,000,000/1, which was more suitable for general than for specialist use. However, it was the first comprehensive topographical atlas ever made of China. The interests of Ting and the survey were reflected in many special plates which provided data on the distribution of languages, aboriginal peoples, mineral resources, and on climate and agricultural production. For twenty years it remained a standard reference, and it went into its fifth revised edition in 1948. See Ting Wen-chiang, Wong Wen-hao, and Tseng Shih-ying, *Chung-hua min-kuo hsin ti-t'u* (New atlas of the Chinese Republic) (Shanghai, Shen pao kuan, 1934). A popular version of the atlas, under the title *Chung-kuo fen-sheng hsin ti-t'u* (New atlas of China's provinces), first appeared in 1933.

10. Hu Shih, *Ting Wen-chiang ti chuan chi*, pp. 83–84. See also Tsiang T'ing-fu, "Wo so chi-te ti Ting Tsai-chün," pp. 139–40.

11. Editorial in *TLPL*, No. 1:2 (May 22, 1932).

12. Ting Wen-chiang, "Chung-kuo cheng-chih ti ch'u-lu" (The road ahead in Chinese politics), *TLPL*, No. 11:6 (July 31, 1932).

13. Hu Shih, *Ting Wen-chiang ti chuan chi*, pp. 83–84.

14. Ting Wen-chiang, "K'ang Jih chiao fei yü chung-yang ti cheng-chü" (Resist Japan, destroy the bandits, and the central government's political situation), *TLPL*, No. 19: 8–10 (Sept. 25, 1932); and "Kuan-yü kuo fang ti ken-pen wen-t'i" (On the basic problem of national defense), *Kuo-wen chou-pao* (Oct. 10, 1934).

15. Ting Wen-chiang, "K'ang Jih ti hsiao-nung yü ch'ing-nien ti tse-jen" (The capacity to resist Japan and youth's responsibility), *TLPL*, No. 37: 2–8 (Feb. 12, 1933).

16. Ting Wen-chiang, "Chia-ju wo shih Chang Hsüeh-liang" (If I were Chang Hsüeh-liang), *TLPL*, No. 13 (Aug. 14, 1932). See also Ting's "Chia-ju wo shih Chiang Chieh-shih" (If I were Chiang Kai-shek), *TLPL*, No. 35 (Jan. 15, 1933).

17. Ting Wen-chiang, "Chi Chang Hsüeh-liang Chiang-chün i-feng kung-kai ti hsin" (An open letter to general Chang Hsüeh-liang), *TLPL*, No. 41: 8–10 (Mar. 12, 1933).

18. The Generalissimo's rather lame explanation was that mistakes in military intelligence had prevented the Kuomintang from accurately judging the timing of the Japanese attack and so sending effective reinforcements to Jehol. Hu Shih and Wong Wen-hao were also present at the interview.

19. Ting Wen-chiang, "Su-o ko-ming wai-chiao shih ti i yeh chi ch'i chiao-hsün (A page in the history of the foreign relations of the Russian revolution and its lessons), *TLPL*, No. 163: 15 (Aug. 11, 1935).

20. Tsiang T'ing-fu, "Wo so chi-te ti Ting Tsai-chün," pp. 141–142.

21. Ting Wen-chiang, "Shih-hsing t'ung-chih ching-chi ti t'iao chien" (Conditions for putting a controlled economy into operation), *TLPL*, No. 108: 18–20 (July 8, 1934).

22. Ting Wen-chiang, "K'o-hsüeh-hua ti chien-she" (Construction made scientific), *TLPL*, No. 151: 10 (May 19, 1935).

23. *TLPL*, No. 51: 3.

24. Ting Wen-chiang, "P'ing-lun Kung-ch'an chu-i ping chung-kao Kung-ch'an tang yüan" (Critique of Communism and counsels to Communists), *TLPL*, No. 51: 5–15 (May 21, 1933).

25. *Ibid.*, p. 6.

26. Ting's views on Marxist thought closely echo those of the British socialist Harold Laski, who was one of Ting's favorite authors. There are strong similarities between Ting's analysis in this article and Laski's presentation in *Communism* (Home University Library, London and New York, 1927). Ting followed Laski in criticizing the labor theory of value

and dialectical materialism as well as in contrasting Marx's theoretical errors with his political insights. Like Laski, he also objected to the use of Marxist dogma to justify a violent insurrectionary movement; and he used historical arguments to show that European capitalist society had not evolved in the way Marx had predicted.

27. The official subsidy available to geologists going to the Washington conference was hardly sufficient for the expenses of global transit and of two months inside Russia itself. In his account of the journey Ting mentioned "several friends" who in 1932 urged him to make the trip and who were later willing to "permit" him to use the 1933 geological congress as an occasion. Further, he reached an understanding with them that his visit was to be solely for the purpose of studying Soviet science. Probably the Nationalist administration, anxious to extend its diplomatic contact with a powerful neighbor, was happy to subsidize an unofficial cultural emissary. See Ting Wen-chiang, "Su-o lü-hsing chi," pp. 439–512. The entire existing report of the trip was first published in 19 installments in the *TLPL*.

28. *Ibid.*, pp. 449–450.

29. *Ibid.*, p. 462.

30. *Ibid.*, p. 467.

31. Ting Wen-chiang, "Tsai-lun min-chu yü tu-ts'ai" (More discussion on democracy and dictatorship), *TLPL*, No. 137: 21–22 (Jan. 27, 1935).

32. Hu Shih, *Ting Wen-chiang ti chuan chi*, pp. 97–98.

33. Ting Wen-chiang, "Wo-ti hsin-yang," pp. 11–12.

34. Ting Wen-chiang, "P'ing-lun Kung-ch'an chu-i ping chung-kao Kung-ch'an tang yüan," pp. 11–14.

35. Ting Wen-chiang, "Wo-ti hsin-yang," pp. 11–12.

36. See especially, Hu Shih, "Ta Ting Tsai-chün hsien-sheng lun min-chu yü tu-ts'ai" (Discussion of democracy and dictatorship: Reply to Mr. Ting Tsai-chün), *TLPL*, No. 133 (Dec. 30, 1934); T'ao Meng-ho (Ming Sheng), "Shuang-chou hsien-tan" (Biweekly conversations), *TLPL*, No. 133 (Dec. 30, 1934); Wu Ching-ch'ao, "Chung-kuo ti cheng-chih wen-t'i" (The problem of China's politics), *TLPL*, No. 134 (Jan. 6, 1935); Ch'en Chih-mai, "Min-chu yü tu-ts'ai ti t'ao-lun" (The debate on democracy and dictatorship), *TLPL*, No. 136 (Jan. 20, 1935).

37. Tsiang T'ing-fu, "Ko-ming yü chuan-chih" (Revolution and autocracy), *TLPL*, No. 80 (Dec. 10, 1933); and "Lun chuan-chih ping ta Hu Shih-chih hsien-sheng" (Discussion of autocracy and a reply to Mr. Hu Shih), *TLPL*, No. 83 (Dec. 31, 1933). For translated selections

from this exchange between Tsiang and Hu, see de Bary, ed., *Sources of Chinese Tradition*, pp. 786–796.

38. See especially Harold J. Laski, *Democracy in Crisis* (Chapel Hill, N. C., 1933); and H. G. Wells, *Democracy Under Revision* (London, 1927).

39. Ting Wen-chiang, "Min-chu cheng-chih yü tu-ts'ai cheng-chih" (Democratic politics and the politics of dictatorship), *TLPL*, No. 133: 6 (Dec. 30, 1934). See also Ting's "Tsai lun min-chu yü tu-ts'ai," pp. 19–22.

40. Ting Wen-chiang, "Min-chu cheng-chih yü tu-ts'ai cheng chih", p. 6.

41. Ting Wen-chiang, "Chung-yang yen-chiu yuan ti shih-ming" (The task of the Academia Sinica), *Tung-fang tsa-chih*, 32.2: 5–8 (Jan. 16, 1935).

42. Ting Wen-chiang, "Kung-kung hsin-yang yü t'ung-i" (Unity and a common faith), *Ta kung pao* (Jan. 14, 1934).

43. Wong, W. H., "V. K. Ting, Biographical Note," *BGSC*, 16: x (1936).

44. In spite of his reservations about the Kuomintang, Ting was coming closer to a position of open cooperation with Chiang Kai-shek as the national emergency deepened, and in his last years the question of whether to accept a post continued to arise from time to time. After Ting died, Fu Ssu-nien wrote that his friend had not planned to serve in government again, but another source claims that in 1936 Chiang was preparing to make Ting minister of railways. See Ch'en Po-chuang, "Chi-nien Ting Tsai-chün hsien-sheng" (Recollections of Mr. Ting Tsai-chün) in Hu Shih, et al., *Ting Wen-chiang che-ko jen*, p. 225.

45. Ting brought a number of independent organizations into the framework of the Academia Sinica structure. The Social Research Institute (*She-hui t'iao-ch'a so*) of Peking was amalgamated with the Academia Sinica's own Social Science Research Institute; the Anthropology Group (*Min-tsu tsu*) was placed under the Academia Sinica Institute of History and Philology and provided with a special subsidy from the China Foundation; the Academia Sinica created a new Institute of Biology in cooperation with the old Metropolitan Museum of Natural History (*Chung-yang tsu-jan li-shih po-wu yüan*), and installed a Textile Laboratory (*Mien fang-chih jan shih-yen-kuan*) as its subsidiary, designed to undertake research into agriculture-based industry. In addition Ting centralized the budgets of the various Academia Sinica institutes so that individual institutes, instead of receiving a fixed percentage of the total budget each year, were financed upon the basis of specific project pro-

posals which would be subsidized in proportion to their merit and urgency. He also reorganized the Academia Sinica endowment fund, reduced the number of members on its executive board, and oversaw the creation of a nationwide Academic Senate (*P'ing-lun hui*). This last group drew its membership from the Academia Sinica and all important universities and independent research institutes, and it was planned as an overall coordinator of research and spokesman for the academic community in public affairs. In practice, however, the senate did little more than gather data on the profession and confer honorary degrees. See Chu Chia-hua, "Ting Wen-chiang yü Chung-yang yen-chiu yüan" (Ting Wen-chiang and the Academia Sinica), *CYYCY*, 3: 126 (December 1956); Ts'ai Yüan P'ei, "Ting Tsai-chün hsien-sheng tui-yü Chung-yang yen-chiu yüan chih kung-hsien" (Mr. Ting Tsai-chün's contributions to the Academia Sinica), *TLPL*, No. 188: 31–32 (Feb. 16, 1936); V. K. Ting, "Modern Science in China," pp. 133–134.

46. Ting Wen-chiang, "Chung-yang yen-chiu yüan chih shih-ming."

47. Li Chi, "Tui-yü Ting Wen-chiang so t'i-ch'ang ti k'o-hsüeh yen-chiu chi tuan hui-i," pp. 158.

48. Hu Shih, *Ting Wen-chiang ti chuan chi*, pp. 80, 111. See also Fu Ssu-nien (Fu Meng-chen), "Ting Wen-chiang i-ko jen-wu," p. 11.

49. Ting Wen-t'ao, p. 46.

BIBLIOGRAPHY

BGSC: Bulletin of the Geological Society of China. Vols. 1–16 (1922–1937).

Biggerstaff, Knight. *The Earliest Modern Government Schools in China.* Ithaca: Cornell University Press, 1961.

Black, Davidson. "The Geological Society and Science in China," *Bulletin of the Geological Society of China.* Vol. 1, No. 1 (March 1922).

Brière, O. *Fifty Years of Chinese Philosophy, 1898–1950.* Tr. Laurence G. Thompson. London: Allen and Unwin, 1956.

British Parliamentary Papers. China, No. 2: 1926. Report of the Advisory Committee Together with Other Documents Respecting the China Indemnity.

Bulletin of the Geological Society of China, see *BGSC.*

Chan, Wing-tsit. "Neo-Confucianism and Chinese Scientific Thought," *Philosophy East and West.* Vol. 6, No. 4 (January 1957).

Chang, Carsun (Chang Chün-mai), *The Development of Neo-Confucian Thought.* 2 vols.; New York: Bookman Associates, 1957–1962.

——*Wang Yang-ming.* Jamaica, New York: St. John's University Press, 1962.

——"Reflections on the Philosophical Controversy in 1923," *Chung Chi Journal,* Vol. 3, No. 1 (November 1963).

Chang Ch'i-yün 張其昀 "Ting Wen-chiang hsien-sheng chu-tso hsi-nien mu-lü" 丁文江先生著作繫年目錄 (Chronological catalogue of Mr. Ting Wen-chiang's writings); *Tu-li p'ing-lun* 獨立評論 No. 188 (Feb. 16, 1936).

Chang Chün-mai (Carsun Chang) 張君勱, ed. *Jen-sheng kuan chih lun-chan* 人生觀之論戰 (The polemic on a view of life). Shanghai, 1923.

————"Tsai lun jen-sheng kuan yü k'o-hsüeh ping ta Ting Tsai-chün" 再論人生觀與科學並答丁在君 (More discussion of a view of life and science, with a reply to Ting Tsai-chün); in *K'o-hsüeh yü jen-sheng kuan*. Shanghai, 1923.

————"Jen-sheng kuan lun-chan chih hui-ku" 人生觀論戰之回顧 (Looking back on the debate over a view of life); *Tung-fang tsa-chih* 東方雜誌, 31: 13: 5–13 (July 1, 1934).

Chang Hung-chao 章鴻釗 "Wo tui-yü Ting Tsai-chün hsien-sheng ti hui-i" 我對於丁在君先生的回憶 (My memories of Mr. Ting Tsai-chün); in *Ti-chih lun-p'ing: Ting Wen-chiang hsien-sheng chi-nien hao* 地質論評：丁文江先生紀念號 (Geological discussions: Issue in commemoration of Ting Wen-chiang); Vol. 1, No. 3 (July 1936).

Chang Yüeh-jui 張越瑞 *Chin jen chuan-chi wen-hsüan* 近人傳記文選 (Selected contemporary biographies). Changsha, 1938.

Ch'en Chih-mai 陳之邁 "Min-chu yü tu-ts'ai ti t'ao-lun" 民主與獨裁的討論 (The debate on democracy and dictatorship); *Tu-li p'ing-lun*, No. 136 (Jan. 20, 1935).

Ch'en Tu-hsiu 陳獨秀 "K'o-hsüeh yü jen-sheng kuan hsü" 國學與人生觀序 (Preface to Science and a View of Life); in *K'o-hsüeh yü jen-sheng kuan*. Shanghai, 1923.

Chesneaux, J. *Le mouvement ouvrier chinois de 1919 à 1927*. Paris, 1962.

Chia I-chün 賈逸君 *Chung-hua min-kuo shih* 中華民國史 (History of the Chinese Republic). Peking, 1930.

Chiang Monlin. *Tides from the West: A Chinese Autobiography*. New Haven: Yale University Press, 1947.

China Yearbook, The, 1926–1927.

China Yearbook, The, 1928.

Chou I-ch'un 周詒春 "Wo so ching-yang ti Ting Tsai-chün hsien-sheng" 我所敬仰的丁在君先生 (My esteemed Mr. Ting Tsai-chün); *Tu-li p'ing-lun*, No. 188 (Feb. 16, 1936).

Chow Tse-tsung. *The May Fourth Movement*. Cambridge: Harvard University Press, 1960.

Chu Chia-hua 朱家驊 "Ting Wen-chiang yü Chung-yang yen-chiu-yüan" 丁文江與中央研究院 (Ting Wen-chiang and the Academia Sinica); *Chung-yang yen-chiu yüan yüan-k'an*, No. 3 (December 1956).

Chu Ching-nung 朱經農 "Tsui-hou i-ko yüeh ti Ting Tsai-chün hsien-sheng" 最後一個月的丁在君先生 (Mr. Ting Tsai-chün's last month); *Tu-li p'ing-lun*, No. 188 (Feb. 16, 1936).

Chu, Coching. "Science in New China," *People's China*. Vol. 1, No. 10 (May 16, 1950).

Chung-kuo k'uang-yeh chi-yao 中國鑛業紀要 (Chinese mining bulletin). Published by the Geological Survey of China. Peking, 1921–.

Chung-yang yen-chiu yüan yüan-k'an, see *CYYCY*.

Chyne, W. Y., ed. *Handbook of Cultural Institutions in China*. Shanghai: Chinese National Committee on Intellectual Cooperation, 1936.

CYYCY: Chung-yang yen-chiu yüan yüan-k'an 中央研究院院刊 (Annals of the Academia Sinica).

Davenport, Charles Benedict. *Heredity in Relation to Eugenics*. New York: Henry Holt and Company, 1915.

Dewey, John. *Reconstruction in Philosophy*. New York: Henry Holt and Co., 1920.

Eucken, Rudolf. *The Problem of Human Life*, tr. Williston H. Hough and W. R. Boyce Gibson. New York: Charles Scribner's Sons, 1910.

————*Life's Basis and Life's Ideal*, tr. Alban G. Widgery. London: Adam & Charles Black, 1911.

Eucken, Rudolf and Carsun Chang. *Das Lebensproblem in China und in Europa*. Leipzig: Quelle and Meyer, 1922.

Franck, Harry A. *Roving Through Southern China*. New York and London: The Century Co., 1925.

Fu Ssu-nien (Fu Meng-chen) 傅斯年 (傅孟真) "Wo so jen-shih ti Ting Wen-chiang hsien-sheng" 我所認識的丁文江先生 (The Ting Wen-chiang I knew); *Tu-li p'ing-lun*, No. 188 (Feb. 16, 1936).

————"Ting Wen-chiang i-ko jen-wu ti chi-p'ien kuang-ts'ai" 丁文江一個人物的幾片光彩 (A few of the glories of Ting Wen-chiang as a personality); *Tu-li p'ing-lun*, No. 189 (Feb. 23, 1936).

Galton, Francis. *Hereditary Genius; An Inquiry into Its Laws and Con-*

sequences. 2nd ed.; London: Watts, 1914 (reprint of the 2nd edition of 1892).

Gillin, Donald G. *Warlord: Yen Hsi-shan in Shansi Province, 1911–1949*. Princeton, N. J.: Princeton University Press, 1967.

Gould, Sydney H., ed. *Sciences in Communist China*. Washington, D. C., 1961.

Grabau, Amadeus W. (Ko-li-p'u) 葛利普 "Ting Wen-chiang hsien-sheng yü Chung-kuo k'o-hsüeh chih fa-chan" 丁文江先生與中國科學之發展 (Mr. Ting Wen-chiang and the development of Chinese science), translated into Chinese by Kao Chen-hsi; *Tu-li p'ing-lun*, No. 188 (Feb. 16, 1936).

Greene, John C. *Darwin and the Modern World View*. Baton Rouge, La.: Louisiana State University Press, 1961.

Huang Chi-ch'ing 黃汲清 "Ting Tsai-chün hsien-sheng tsai ti-chih hsüeh shang ti kung-tso" 丁在君先生在地質學上的工作 (Mr. Ting Tsai-chün's work in geology); *Tu-li p'ing-lun*, No. 188 (Feb. 16, 1936).

Hsiang tao 嚮導 (May-December, 1926).

Hsieh Chiao-min. "Hsia-ke Hsu: Pioneer of Modern Geography in China," *Annals of the Association of American Geographers*, Vol. 48 (March 1958).

Hsin T'ang-shu. "Hsia-ke Hsu: A Great Geographer in China," *Association of American Geographers*, No. 45 (1945).

Hu Chen-hsing 胡振興 "Shui sung chi Ting Wen-chiang hsien-sheng wu-ch'ien yüan?" 誰送給丁文江先生五千元 (Who gave Mr. Ting Wen-chiang five thousand yüan?); *Tu-li p'ing-lun*, No. 193 (Mar. 22, 1936).

Hu Shih 胡適 *The Development of the Logical Method in Ancient China*. Shanghai: The Oriental Book Co., 1922.

———"Kuan-yü 'Wo-men ti cheng-chih chu-chang' ti t'ao-lun" 關於 '我們的政治主張' 的討論 (On criticisms of "Our Political Proposals"); *Nu-li chou pao* 努力週報 No. 4 (May 28, 1922).

———"K'o-hsüeh yü jen-sheng kuan hsü" 科學與人生觀序 (Preface to Science and a View of Life); in *K'o-hsüeh yü jen-sheng kuan*. Shanghai, 1923.

———"K'o-hsüeh ti ku-shih-chia Ts'ui Shu" 科學的古史家崔述 (Ts'ui Shu, a scientific historian); *Kuo-hsüeh chi-k'an* 國學季刊 (Sinological journal), Vol. 1, No. 2 (April 1923).

———"Tai Tung-yüan ti che-hsüeh" 戴東原的哲學 (The philosophy of

Tai Tung-yüan); *Kuo-hsüeh chi-k'an* 國學季刊 Vol. 2, No. 1 (December 1925).

———*Hu Shih wen-ts'un* 胡適文存 (Collected writings of Hu Shih), Ser. 3, 1930.

———"What I believe," *Forum Magazine*. February 1931.

———*The Chinese Renaissance*. Chicago: The University of Chicago Press, 1934.

———"Ta Ting Tsai-chün hsien-sheng lun min-chu yü tu-ts'ai" 答丁在君先生論民主與獨裁 (Discussion of democracy and dictatorship: Reply to Mr. Ting Tsai-chün); *Tu-li p'ing-lun*, No. 133 (Dec. 30, 1934).

———"Ting Tsai-chün che-ko jen" 丁在君這個人 (This man, Ting Tsai-chün); *Tu-li p'ing-lun*, No. 188 (Feb. 16, 1936).

———"Ting Wen-chiang ti chuan chi" 丁文江的傳記 (Biography of Ting Wen-chiang); *Chung-yang yen-chiu yüan yüan-k'an*, No. 3 (December 1956).

———"The Scientific Spirit and Method in Chinese Philosophy," *Philosophy East and West*, Vol. 9, Nos. 1 and 2 (April and July 1959).

———"Cheng tzu-yu ti hsüan-yen" 爭自由的宣言 (Manifesto of struggle for freedom); *Tung-fang tsa-chih* 東方雜誌 (Eastern miscellany), Vol. 17, No. 16 (Aug. 25, 1920).

———"Wo-men ti cheng-chih chu-chang" 我們的政治主張 (Our political proposals); *Nu-li chou-pao*, No. 2 (May 14, 1922).

———et al. *Ting Wen-chiang che-ko jen* 丁文江這個人 (This man, Ting Wen-chiang), Vol. 21 of *Chuan-chi wen-hsüeh ts'ung-shu* 傳記文學叢書 (Library of biographical literature). Taipei, 1968.

Huxley, Thomas. *Collected Works*. Authorized edition; 9 vols.; New York: D. Appleton & Co., 1904–1920.

Irvine, William. *Apes, Angels and Victorians: Darwin, Huxley and Evolution*. New York: McGraw-Hill, 1955.

Israel, John. *Student Nationalism in China, 1927–1936*. Stanford, California: Hoover Institute, 1966.

Jevons, W. Stanley. *The Principles of Science; A Treatise on Logic and Scientific Method*. London and New York: The Macmillan Co., 1920. Reprinted from the 2nd edition of 1877.

Ju Sung 如松 "P'ing yu-sheng-hsüeh yü huan-ching-lun ti lun-cheng" 評優生學與環境論的論爭 (Critique of the debate on eugenics and environment); *Erh-shih shih-chi* 二十世紀 (Twentieth century) (Feb. 1, 1935).

Kao Chen-hsi 高振西 "Tso chiao-shih ti Ting Wen-chiang hsien-sheng." 做教師的丁文江先生 (Mr. Ting Wen-chiang the teacher); *Tu-li p'ing-lun.* No. 188 (Feb. 16, 1936).

KHY: K'o-hsüeh yü jen-sheng kuan. 科學與人生觀 (Science and a View of Life), prefaces by Hu Shih and Ch'en Tu-hsiu. Shanghai, 1923.

K'o-hsüeh yü jen-sheng kuan, see *KHY.*

Kotenev, A. M. *Shanghai: Its Mixed Court and Council.* Shanghai: North-China Daily News and Herald, Ltd., 1925.

————*Shanghai: Its Municipality and the Chinese.* Shanghai: North-China Daily News and Herald, Ltd., 1927.

Kropotkin, Petr Alekseevich. *Mutual Aid: A Factor of Evolution.* Harmondsworth, Middlesex, England: Penguin Books, Ltd., 1939.

Ku Chieh-kang. *Autobiography of a Chinese Historian: Being the Preface to a Symposium on Ancient Chinese History,* tr. Arthur W. Hummel. Leyden: E. J. Brill, Ltd., 1931.

Kuo Mo-jo. "A New Stage in Chinese Science," *People's China,* No. 14 (1955).

Kuo-wen chou-pao 國聞週報 (May–December 1926).

Kwok, D. W. T. "Wu Chih-hui and Scientism," *Tsing Hua Journal of Chinese Studies.* New Ser. 3, No. 1 (May 1962).

————*Scientism in Chinese Thought, 1900–1950.* New Haven: Yale University Press, 1965.

Laski, H. J. *Communism* (Home University Library; New York: H. Holt and Company, 1927).

————*Democracy in Crisis.* Chapel Hill: The University of North Carolina Press, 1933.

Levenson, Joseph R. *Liang Ch'i-ch'ao and the Mind of Modern China.* Cambridge, Mass.: Harvard University Press, 1953.

————*Confucian China and Its Modern Fate.* 3 vols.; Berkeley: University of California Press, 1958–1965.

Li Chi 李濟 "Tui-yü Ting Wen-chiang so t'i-ch'ang ti k'o-hsüeh yen-chiu chi-tuan hui-i" 對於丁文江所提倡的科學研究幾段回憶 (Some recollections of scientific research pioneered by Ting Wen-chiang); *Chung-yang yen-chiu-yüan yüan-k'an,* No. 3 (December 1956).

Li Chien-nung. *The Political History of China, 1840–1928.* tr. Ssu-yu Teng and Jeremy Ingalls. Princeton: D. Van Nostrand & Co., 1956.

Li Hsüeh-ch'ing 李學清 "Chui-nien Ting shih Tsai-chün hsien-sheng" 追念丁師在君先生 (Recollections of Professor Ting Tsai-chün);

in *Ti-chih lun-p'ing: Ting Wen-chiang hsien-sheng chi-nien hao* 地質論評：丁文江先生紀念號 Vol. 1, No. 3 (July 1936).

Li Ssu-kuang. "Science Serves the People," *People's China*, Vol. 4, No. 4 (Aug. 16, 1951).

Li Tsu-hung (Li I-shih) 李祖鴻（李毅士） "Liu-hsüeh shih-tai ti Ting Tsai-chün" 留學時代的丁在君 (Ting Tsai-chün as a student abroad); *Tu-li p'ing-lun*, No. 208 (July 5, 1936).

Liang Ch'i-ch'ao 梁啓超 "Hsin min shuo"; 新民説 (A new people); in Vol. 19 of *Yin-ping-shih ho-chi* 飲冰室合集 (Collected works), 40 vols.; Shanghai, 1936.

———*Intellectual Trends in the Ch'ing Period*, tr. Immanuel C. Y. Hsü. Cambridge, Mass.: Harvard University Press, 1959.

———"O yu hsin-ying lu chieh-lu" 歐遊心影錄節錄 (Reflections on a trip to Europe); in *Yin-ping-shih ho-chi*, Vol. 21.

———"Chung-kuo li-shih yen-chiu fa" 中國歷史研究法 (Method in the study of Chinese history); in *Yin-ping-shih ho-chi*, Vol. 32.

———"K'o-hsüeh ching-shen yü Tung-Hsi wen-hua" 科學精神與東西文化 (The Scientific spirit and Eastern and Western culture); in *Liang Jen-kung hsüeh-shu chiang-yen chi* 梁任公學術講演集 (Liang Ch'i-ch'ao's academic writings), Shanghai, 1922.

Liang Sou-ming 梁漱溟 *Tung hsi wen-hua chi ch'i che-hsüeh* 東西文化及其哲學 (Eastern and Western culture and their philosophies). Shanghai, 1921.

Lin, Mousheng Hsitien. "Recent Intellectual Movements in China," *China Institute Bulletin*, Vol. 3, Nos. 1 and 3 (October and December 1938).

Ling Hung-hsün 凌鴻勛 "Tao Ting Tsai-chün hsien-sheng" 悼丁在君先生 (In mourning for Mr. Ting Tsai-chün); *Tu-li p'ing-lun*, No. 188 (Feb. 16, 1936).

Liu, F. F. *A Military History of Modern China*. Princeton: Princeton University Press, 1956.

Lo Chia-lun 羅家倫 "Hsien-tai hsüeh-jen Ting Tsai-chün ti i-chüeh" 現代學人丁在君的一角 (On the contemporary scholar Ting Tsai-chün); *Chung-yang yen-chiu yüan yüan-k'an*, No. 3 (December 1956).

Mach, Ernst. *The Analysis of Sensations*. Translated from the 1st German edition by C. M. Williams; revised and supplemented from the 5th German edition by Sidney Waterlow. Chicago and London: Open Court Publishing Co., 1914.

Meng Hui-min 孟憲民. "Yunnan Ko-chiu ti-chih shu lüeh" 雲南箇舊地質述略 (Sketch of the geology of Kokiu in Yunnan); *Ti-chih lun-p'ing: Ting Wen-chiang hsien-sheng chi-nien hao*, Vol. 1, No. 3 (July 1936).

Needham, Joseph. *Science and Civilization in China*. Cambridge, England: Cambridge University Press, 1954–.
North China Herald (May-December 1926).
Nu-li chou-pao 努力週報 (Endeavor magazine), Nos. 1–75.

Peake, Cyrus H. "Some Aspects of the Introduction of Modern Science into China," *Isis*, Vol. 22, No. 63 (December 1934).
Pearson, Karl. *Nature and Nurture: the Problem of the Future*. London: University of London, Galton Laboratory for National Eugenics, 1910.
——— *The Scope and Importance to the State of the Science of National Eugenics*. 3rd ed.; London: Cambridge University Press, 1911.
——— *The Problem of Practical Eugenics*. London: Cambridge University Press, 1912.
———*Darwinism, Medical Progress and Eugenics*. London: University of London, Galton Laboratory for National Eugenics, 1912.
——— *The Grammar of Science* (Everyman's Library; London: J. M. Dent & Sons, Ltd., 1937).

Richthofen, Ferdinand von. *China: Ergebnisse einer Reisen und darauf gregründeter Studien*. 5 vols.; Berlin: D. Reimer, 1877–1912.
Russell, Bertrand. *The Problem of China*. London: Allen and Unwin, 1922.

Sakai, R. K. "Ts'ai Yüan-p'ei as a Synthesizer of Western and Chinese Thought," *Papers on China*, 3: 170–192. Harvard University, East Asian Research Center, 1949.
Schwartz, Benjamin. "Ch'en Tu-hsiu and the Acceptance of the Modern West," *Journal of the History of Ideas*. Vol. 12, No. 1 (January 1951).
———*In Search of Wealth and Power: Yen Fu and the West*. Cambridge: Harvard University Press, 1964.
Sun E-tu Zen and Sun Shiou-chuan, tr. *T'ien kung k'ai wu: Chinese Technology in the Seventeenth Century*. University Park and London: University of Pennsylvania Press, 1966.

282

Sung Ying-hsing 宋應星. *T'ien kung k'ai wu* 天工開物 (The creations of nature and human labor), ed. T'ao Hsiang 陶湘 with a critique and biography of the author by Ting Wen-chiang. Shanghai, 1929.

T'an Ssu-t'ung 譚嗣同 *Jen hsüeh* 仁學 (The study of jen). Peking, 1958.

T'ang Chung 湯中 "Tui-yü Ting Tsai-chün hsien-sheng ti hui-i" 對於丁在君先生的回憶 (Recollections of Mr. Ting Tsai-chün); *Tu-li p'ing-lun*. No. 211 (July 26, 1936).

T'ao Meng-ho (Ming Sheng) 陶孟和 (明生). "Shuang-chou hsien-t'an" 雙週閑談 (Biweekly conversations); *Tu-li p'ing-lun*, No. 133 (Dec. 30, 1934).

——"Chui-i Tsai-chün" 追憶在君 (Recollections of Tsai-chün); *Tu-li p'ing-lun*. No. 188 (Feb. 16, 1936).

Teilhard de Chardin, Pierre. *Letters From a Traveller*. Tr. Bernard Wall. London: Collins, 1962.

Thomson, J. Arthur. *Introduction to Science*. New York: Henry Holt and Company, 1911.

Ti-chih hui-pao 地質彙報 (Bulletin of the Geological Survey of China). Published by the Geological Survey of China, July 1919–.

Ting Chang Tzu-san 丁張紫珊 "Tao Tsai-chün erh-ko" 悼在君二哥 (In mourning for second brother Tsai-chün); *Kuo-wen chou-pao* 國聞週報 Vol. 13, No. 4.

Ting, V. K. (Ting Wen-chiang), "Chinese Students," *Westminster Review*, Vol. 169, No. 1 (January 1908).

——"Tungchwanfu, Yunnan, Copper Mines," *Far Eastern Review*, Vol. 12, No. 5 (October 1915).

——"The Coal Resources of China," *Far Eastern Review*, Vol. 13, No. 1 (June 1916).

——"China's Mineral Resources," *Far Eastern Review*, Vol. 13, No. 14 (July 1917).

——"Mining Legislation and Development in China," *Far Eastern Review*, Vol. 13, No. 14 (July 1917).

——"On Hsu Hsia-k'o (1586–1641), Explorer and Geographer," *New China Review*, Vol. 3, No. 5 (October 1921).

——"Presidential Address," *Bulletin of the Geological Society of China*, Vol. 3, No. 1 (1924).

——"The Greater Shanghai Municipality," *North China Herald* (Oct. 16, 1926).

——"Biographical Note: A. W. Grabau," *Bulletin of the Geological*

Society of China, Vol. 10 (1931).

———"How China Acquired her Civilization," in *Symposium on Chinese Culture*, ed. Sophia H. Chen Zen. Shanghai: China Institute of Pacific Relations, 1931.

———"Professor Granet's 'La civilization chinoise,'" *Chinese Social and Political Science Review* (July 1931).

———"Reconnaissance of a Railway Line from Chungking to Kwang-chowwan," *Far Eastern Review* (June 1932).

———"If I Were Chang Hsüeh-liang," *People's Tribune*. Vol. 3, New Ser., No. 5 (Oct. 1, 1932).

———"On the Japanese Invasion, the Communist Campaign and the Kuomintang," *People's Tribune*, Vol. 3, New Ser., No. 6 (Oct. 16, 1932).

———"Notizen von einer gemächlichen Fahrt in Südchina," *Anthropos*, Vol. 29 (1934).

———"Research in China: the Academia Sinica," *Nature*, Vol. 136 (Aug. 10, 1935).

———"Modern Science in China," *Asia*, Vol. 36, No. 2 (February 1936).

———tr. "Things Produced by the Works of Nature," *Mariner's Mirror: The Quarterly Journal of the Society for Nautical Research*. Vol. 11, No. 3 (July 1935).

Ting Wen-chiang 丁文江 (V. K. Ting). *Tung-wu-hsüeh chiao-k'o-shu* 動物學教科書 (Textbook in zoology). Shanghai, 1914.

———"Che-ssu-hsüeh yü p'u-tieh" 哲嗣學與譜牒 (Eugenics and genealogy); *Kai-tsao* 改造 (Reconstruction), Vol. 3, Nos. 4, 5, and 6 (1919).

———"Wu-shih-nien lai Chung-kuo chih k'uang-yeh" 五十年來中國之礦業 (Mining in China during the past fifty years); *Tsui-chin wu-shih-nien* 最近五十年 (The past fifty years). Shen Pao anniversary volume. Shanghai, 1922.

———"Li-shih jen-wu yü ti-li chih kuan-hsi" 歷史人物與地理之關係 (The relation between geography and famous men in history); *K'o-hsüeh* 科學 (February 1922).

———(Tsung Yen) 宗淹 "Chün-shih t'iao-ch'a" 軍事調查 (Research into military affairs); *Nu-li chou-pao*, No. 1 (May 7, 1922).

———(Tsung Yen). "Feng-Chih chan-cheng ti chen-hsiang" 奉直戰爭的真像 (An accurate account of the Fengtien-Chihli war); *Nu-li chou-pao*, No. 3 (May 21, 1922).

———(Tsung Yen). "Kuang-tung chün-tui ti kai-lüeh" 關東軍隊的

概略 (Sketch of Kwangtung's army); *Nu-li chou-pao*, No. 5 (June 4, 1922).

————(Tsung Yen). "Ta kuan-yü 'Wo-men ti cheng-chih chu-chang'" 答關於我們的政治主張 (Reply concerning "Our Political Proposals"); *Nu-li chou-pao*, No. 6 and No. 7 (June 11 and June 18, 1922).

————(Tsung Yen). "Chung-kao chiu kuo-hui i-yüan" 忠告舊國會議員 (Counsels to the deputies of the old parliament); *Nu-li chou-pao*, No. 9 (July 2, 1922).

————"Ts'ai-ping chi-hua ti t'ao-lun" 裁兵計劃的討論 (Criticism of plans for disarmament); *Nu-li chou-pao*, No. 14 (Aug. 6, 1922).

————"Hunan chün-tui kai-lüeh" 湖南軍隊概略 (Sketch of Hunan's army); *Nu-li chou-pao*, No. 19 (Sept. 10, 1922).

————"Shan-hai-kuan wai lü-hsing chien-wen lu" 山海關外旅行見聞錄 (Eyewitness report of a journey beyond Shanhaikuan); *Nu-li chou-pao*, No. 28 (Nov. 12, 1922).

————"I-ko wai-kuo p'eng-yu tui-yü i-ko liu-hsüeh-sheng ti chung-kao" 一個外國朋友對於一個留學生的忠告 (Counsels of a foreign friend to a Chinese graduate from abroad); *Nu-li chou-pao*, No. 42 (Mar. 23, 1923).

————"Lan-yin-ho p'an ti pei-ch'ü" 蘭因河畔的悲劇 (The tragedy of the Rhine frontier); *Nu-li chou-pao*, No. 47 (Apr. 8, 1923).

————"Shao-shu jen ti tse-jen" 少數人的責任 (The responsibility of a few men); *Nu-li chou-pao*, No. 67 (Oct. 6, 1923).

————"Hsüan-hsüeh yü k'o-hsüeh" 玄學與科學 (Metaphysics and science); in *K'o-hsüeh yü jen-sheng kuan.* Shanghai, 1923.

————"Hsüan-hsüeh yü k'o-hsüeh: Ta Chang Chün-mai" 玄學與科學 —答張君勱 (Metaphysics and science: A reply to Chang Chün-mai); in *K'o-hsüeh yü jen-sheng kuan.* Shanghai, 1923.

————"Hsüan-hsüeh yü k'o-hsüeh ti t'ao-lun ti yü-hsing" 玄學與科學的 討論的餘興 (Envoi to the debate on science and metaphysics); in *K'o-hsüeh yü jen-sheng kuan.* Shanghai, 1923.

————"Hsü Hsia-k'o yu-chi" 徐霞客遊記 (The travels of Hsu Hsia-k'o); *Chung-yang yen-chiu yüan yüan-k'an*, No. 3 (December 1956). Originally appeared in *Hsiao-shuo yüeh-pao* 小說月報 (Short story monthly; special supplement on Chinese culture), Vol. 17 (1926).

————*Min-kuo chün-shih chin-chi* 民國軍事近紀 (Notes on republican military affairs). Shanghai, 1926.

————*Chung-kuo kuan-pan k'uang-yeh shih lüeh* 中國官辦鑛業史略 (Outline history of Chinese official managed mines). Peking, 1928.

———*Wai-tzu k'uang-yeh shih tzu-liao* 外資鑛業史資料 (Materials for a history of foreign capitalized mines). Peking, 1929.

———"Ch'üan-yang pei-tz'u yü Jih-pen cheng-chü ti ch'ien-t'u" 犬養被刺與日本政局的前途 (The assassination of Inukai and the future of the Japanese political situation); *Tu-li p'ing-lun*, No. 1 (May 22, 1932).

———"Jih-pen ti hsin nei-ko" 日本的新内閣 (Japan's new cabinet); *Tu-li p'ing-lun*, No. 2 (May 29, 1932).

———"Jih-pen ti ts'ai-cheng" 日本的財政 (Japan's finances); *Tu-li p'ing-lun*, No. 2 (May 29, 1932).

———"So-wei Pei-ping ko ta-hsüeh ho-li-hua ti yün-tung" 所謂北平各大學合理化的運動 (The so-called movement to integrate all Peking universities); *Tu-li p'ing-lun*, No. 3 (June 5, 1932).

———"So-wei 'chiao fei' wen-t'i" 所謂'剿匪'問題 (The so-called problem of 'destroying banditry'); *Tu-li p'ing-lun*, No. 6 (June 26, 1932).

———"Chung-kuo cheng-chih ti ch'u-lu" 中國政治的出路 (The road ahead in Chinese politics); *Tu-li p'ing lun*, No. 11 (July 31, 1932).

———"Chia-ju wo shih Chang Hsüeh-liang" 假如我是張學良 (If I were Chang Hsüeh-liang); *Tu-li p'ing-lun*, No. 13 (Aug. 14, 1932).

———"K'ang Jih chiao fei yü chung-yang ti cheng-chü" 抗日剿匪與中央的政局 (Resist Japan, destroy the bandits and the central government's political situation); *Tu-li p'ing-lun*, No. 19 (Sept. 25, 1932).

———"Tzu-sha" 自殺 (Suicide); *Tu-li p'ing-lun*, No. 23 (Oct. 23, 1932).

———"Fei-chih nei-chan ti yün-tung" 廢止内戰的運動 (The movement to end civil war); *Tu-li p'ing-lun*, No. 25 (Nov. 6, 1932).

———"Jih-pen ti ts'ai-cheng" 日本的財政 (Japan's finances); *Tu-li p'ing-lun*, No. 30 (Dec. 11, 1932).

———"Chia-ju wo shih Chiang Chieh-shih" 假如我是蔣介石 (If I were Chiang Kai-shek); *Tu-li p'ing-lun*, No. 35 (Jan. 15, 1933).

———"K'ang Jih ti hsiao-nung yü ch'ing-nien ti tse-jen" 抗日的效能與青年的責任 (The capacity to resist Japan and youth's responsibility); *Tu-li p'ing-lun*, No. 37 (Feb. 12, 1933).

———"Wo so chih-tao ti Chu Ch'ing-lan Chiang-chün" 我所知道的朱慶瀾將軍 (The General Chu Ch'ing-lan I know); *Tu-li p'ing-lun*, No. 39 (Feb. 26, 1933).

———"Chi Chang Hsüeh-liang Chiang-chün i feng kung-k'ai ti hsin" 給張學良將軍一封公開的信 (An open letter to General Chang

Hsüeh-liang); *Tu-li p'ing-lun*, No. 41 (Mar. 12, 1933).

—————"P'ing-lun Kung-ch'an chu-i ping chung-kao Kung-ch'an tang-yüan" 評論共產主義並忠告共產黨員 (Critique of Communism and counsels to Communists); *Tu-li p'ing-lun*, No. 51 (May 21, 1933).

—————"Kung-kung hsin-yang yü t'ung-i" 公共信仰與統一 (Unity and a common faith); *Ta kung pao* (Jan. 14, 1934).

—————"Wo-ti hsin-yang" 我的信仰 (My beliefs); *Tu-li p'ing-lun*, No. 100 (May 30, 1934).

—————"Wo so chih-tao ti Weng Jung-ni" 我所知道的翁詠霓 (The Weng Jung-ni [Wong Wen-hao] I know); *Tu-li p'ing-lun*, No. 97 (Apr. 22, 1934).

—————"Shih-hsing t'ung-chih ching-chi ti t'iao-chien" 實行統制經濟的條件 (Conditions for putting a controlled economy into operation); *Tu-li p'ing-lun*, No. 108 (July 8, 1934).

—————"Yin ch'u-k'ou cheng-shui i-hou" 銀出口徵稅以後 (After the silver export tax); *Kuo-wen chou-pao* (Sept. 11, 1934).

—————"Kuan-yü kuo-fang ti ken-pen wen-t'i" 關於國防的根本問題 (On the basic problem of national defense); *Kuo-wen chou-pao* (Oct. 10, 1934).

—————"Min-chu cheng-chih yü tu-ts'ai cheng-chih" 民主政治與獨裁政治 (Democratic politics and dictatorship politics); *Tu-li p'ing-lun*, No. 133 (Dec. 30, 1934).

—————"Chung-yang yen-chiu yüan chih shih-ming" 中央研究院之使命 (The task of the Academia Sinica); *Tung-fang tsa-chih*. Vol. 32, No. 2 (Jan. 16, 1935).

—————"Tsai lun min-chu yü tu-ts'ai" 再論民主與獨裁 (More discussion of democracy and dictatorship); *Tu-li p'ing-lun*, No. 137 (Jan. 27, 1935).

—————"Hsien-tsai Chung-kuo ti chung-nien yü ch'ing-nien" 現在中國的中年與青年 (The middle-aged and the young in China today); *Tu-li p'ing-lun*, No. 144 (Mar. 31, 1935).

—————"K'o-hsüeh-hua ti chien-she" 科學化的建設 (Construction made scientific); *Tu-li p'ing-lun*, No. 151 (May 19, 1935).

—————"Su-o ko-ming wai-chiao shih ti i-yeh chi ch'i chiao-hsün" 蘇俄革命外交史的一頁及其教訓 (A page in the history of the foreign relations of the Russian Revolution and its lessons); *Tu-li p'ing-lun*, No. 163 (Aug. 11, 1935).

—————*Ts'uan wen ts'ung k'e* 爨文叢刻 (Collection of inscriptions in the Ts'uan tongue). Academia Sinica Institute of History and

Philology, Monograph No. 11. Shanghai, 1936.

——"Ch'ien min-yao" 黔民謠 (Ballad of Kweichow); *Tu-li p'ing-lun,* No. 196 (Apr. 12, 1936).

——*Ting-Wen-chiang hsien-sheng ti-chih t'iao-ch'a pao-kao* 丁文江先生地質調查報告 (Geological reports of Dr. V. K. Ting). Nanking: Geological Survey, 1947.

——"Ch'ung yin 'T'ien kung k'ai wu' chüan pa" 重印天工開物卷跋 (Appendix on the reprinting of the "T'ien kung k'ai wu"). Reprinted in *Chung-yang yen-chiu yüan yüan-k'an,* No. 3 (December 1956).

——"Su-o lü-hsing chi" 蘇俄旅行記 (Record of a journey to the Soviet Union). Reprinted in *Chung-yang yen-chiu yüan yüan-k'an,* No. 3 (December 1956).

——"Feng-hsin Sung Chang-keng hsien-sheng chüan" 奉新宋長庚先生傳 (Biography of Mr. Sung Chang-keng of Feng-hsin). Reprinted in *Chung-yang yen-chiu yüan yüan-k'an,* No. 3 (December 1956).

——"Man-yu san-chi" 漫遊散記 (Miscellaneous travels). Reprinted in *Chung-yang yen-chiu yüan yüan-k'an,* No. 3 (December 1956).

——and Tseng Shih-ying 曾世英 "Ch'uan-Kuang t'ieh-tao lu-hsien ch'u-k'an pao-kao" 川廣鐵道路線初勘報告 (Report on a preliminary investigation of a railroad from Szechuan to Kwangtung); *Ti-chih chuan-pao* 地質專報 (Geological memoirs), Ser. B, No. 4 (November 1931). Summary in English.

——Wong Wen-hao and Tseng Shih-ying *Chung-kuo fen-sheng hsin-t'u* 中國分省新圖 (New atlas of China's provinces). Shanghai, 1933.

——Wong Wen-hao and Tseng Shih-ying *Chung-hua min-kuo hsin ti-t'u* 中華民國新地圖 (New atlas of the Chinese Republic). Shanghai, 1934.

——comp. *Hsü Hsia-k'o yu-chi* 徐霞客遊記 (The travels of Hsü Hsia-k'o). 3 vols.; Shanghai, 1928. Includes a "Hsü Hsia-k'o nien-p'u" 徐霞客年譜 (Chronological biography of Hsü Hsia-k'o), and a volume of maps.

——comp. *Liang Jen-kung hsien-sheng nien-p'u ch'ang-pien ch'u-kao* 梁任公先生年譜長編初稿 (Draft materials for a chronological biography of Mr. Liang Jen-kung [Liang Ch'i-ch'ao]). Preface by Ting Wen-yüan. Taipei, 1958.

Ting Wen-chih 丁文治 "Wo-ti erh-ko Wen-chiang" 我的二哥文江 (My second brother Wen-chiang); *Tu-li p'ing-lun,* No. 188 (Feb. 16, 1936).

Ting Wen-t'ao 丁文濤 "Wang ti Tsai-chün t'ung-nien i-shih chui-i lu" 亡弟在君童年軼事追憶錄 (Recollected anecdotes of the youth of my lost brother Tsai-chün); *Tu-li p'ing-lun*, No. 188 (Feb. 16, 1936).

Ting Wen-yüan 丁文淵 "Wen-chiang erh-ko chiao-hsün wo ti ku-shih" 文江二哥教訓我的故事 (The story of what second brother Wen-chiang taught me); *Jo feng* 熱風 (Desert wind), No. 22 (Aug. 1, 1954).

——et al. *Chi-nien Ting Wen-yüan hsien-sheng* 紀念丁文淵先生 (In commemoration of Mr. Ting Wen-yüan). Hongkong, 1963.

TLPL: Tu-li p'ing-lun 獨立評論 (Independent Critic).

Toulmin, Stephen. *The Philosophy of Science: An Introduction*. London and New York: Hutchinson's University Library, 1953.

Ts'ai Yüan-p'ei 蔡元培 "Wo tsai Pei-ching Ta-hsüeh ti ching-li" 我在北京大學的經歷 (My presidency of Peking University); *Tungfang tsa-chih*, Vol. 31, No. 1 (Feb. 14, 1934).

——"Ting Tsai-chün hsien-sheng tui-yü Chung-yang yen-chiu yüan chih kung-hsien" 丁在君先生對於中央研究院之貢獻 (Mr. Ting Tsai-chün's contributions to the Academia Sinica); *Tu-li p'ing-lun*, No. 188 (Feb. 16, 1936).

Tsiang T'ing-fu (Chiang T'ing-fu) 蔣廷黻 "Wo so chi-te ti Ting Tsai-chün" 我所記得的丁在君 (The Ting Tsai-chün I remember); *Chung-yang yen-chiu yüan yüan-k'an*, No. 3 (December 1956).

——"Ko-ming yü chuan-chih" 革命與專制 (Revolution and autocracy); *Tu-li p'ing-lun*, No. 80 (Dec. 10, 1933).

——"Lun chuan-chih ping ta Hu Shih-chih hsien-sheng" 論專制並答胡適之先生 (Discussion of autocracy and a reply to Mr. Hu Shih); *Tu-li p'ing-lun*, No. 83 (Dec. 31, 1933).

Tu-li p'ing-lun, see *TLPL*.

Tung Hsien-kuang (Hollington Tong) 董顯光 "Wo ho Tsai-chün" 我和在君 (Tsai-chün and I); *Chung-yang yen-chiu yüan yüan-k'an*, No. 3 (December 1956).

Tung Tso-pin 董作賓 "Kuan-yü Ting Wen-chiang hsien-sheng ti 'Ts'uan wen ts'ung k'e' chia pien" 關於丁文江先生的爨文叢刻甲編 (Concerning Mr. Ting Wen-chiang's first edition of "Inscriptions in the Ts'uan Tongue"); *Chung-yang yen-chiu yüan yüan-k'an*, No. 3 (December 1956).

Wang Ching-chun. *Japan's Continental Adventure*. New York: Macmillan Co., 1941.

Wang Ching-hsi 汪敬熙 "Ting Tsai-chün hsien-sheng" 丁在君先生 (Mr. Ting Tsai-chün); *Tu-li p'ing-lun*, No. 188 (Feb. 16, 1936).

Wang, Y. C. J. *Chinese Intellectuals and the West*. Chapel Hill, N. C.: University of North Carolina Press, 1966.

Wells, H. G. *Democracy under Revision*. London: Leonard and Virginia Woolf, 1927.

Wong, W. H. (Wong Wen-hao). "Richthofen and Geological Work in China," *Bulletin of the Geological Society of China*, Vol. 12, No. 3 (1933).

——"V. K. Ting, Scientist and Patriot," *Bulletin of the Geological Society of China*. Vol. 15, No. 1 (March 1936).

——"V. K. Ting, Biographical Note," *Bulletin of the Geological Society of China*. Vol. 16 (1936–1937).

Wong Wen-hao 翁文灝 (W. H. Wong). "Tui-yü Ting Tsai-chün hsien-sheng ti chui-i" 對於丁在君先生的追憶 (Recollections of Mr. Ting Tsai-chün); *Tu-li p'ing-lun*, No. 188 (Feb. 16, 1936).

Wu Chih-hui 吳稚暉 "Chen yang-pa-ku-hua chih li-hsüeh" 箴洋八股化之理學 (A warning against neo-Confucian philosophy in Western *pa-ku* disguises); in *K'o-hsüeh yü jen-sheng kuan*. Shanghai, 1923.

——"I-ko hsin hsin-yang ti yü-chou kuan yü jen-sheng kuan" 一個新信仰的宇宙觀與人生觀 (A new belief and its view of life and the universe); in *K'o-hsüeh yü jen-sheng kuan*, Shanghai, 1923.

Wu Ching-ch'ao 吳景超 "Chung-kuo ti cheng-chih wen-t'i" 中國的政治問題 (The problem of China's politics); *Tu-li p'ing-lun*, No. 134 (Jan. 6, 1935).

Wu Ting-liang 吳定良 "Ting Tsai-chün hsien-sheng tui-yü jen-lei hsüeh chih kung-hsien" 丁在君先生對於人類學之貢獻 (Mr. Ting Tsai-chün's contribution to anthropology); *Tu-li p'ing-lun*, No. 188 (Feb. 16, 1936).

Yin Tsan-hsün 尹贊勛. "Yünnan ti-chih yen-chiu ti chin chan" 雲南地質研究的進展 (Recent developments in the geology of Yunnan); *Ti-chih lun-p'ing: Ting Wen-chiang hsien-sheng chi-nien hao*, Vol. 1, No. 3 (July 1936).

Yüan Wei-chou 阮維周 "Ting Tsai-chün hsien-sheng tsai ti-chih hsüeh shang chih kung-hsien" 丁在君先生在地質學上之貢獻 (Mr. Ting Tsai-chün's contribution to geology); *Chung-yang yen-chiu yüan yüan-k'an*, No. 3 (December 1956).

Zen, Sophia H. Chen, ed. *Symposium on Chinese Culture*. Shanghai: China Institute of Pacific Relations, 1931.

GLOSSARY

Chang Chien 張謇

Chang Chün-mai (Chang Chia-sen) 張君勱 (張嘉森)

Chang Hsüeh-liang 張學良

Chang Hung-chao 章鴻釗

Chang I-ou 張軼歐

Chang Tso-lin 張作霖

Chang Tsung-ch'ang 張宗昌

Chang Wei-tz'u 張慰慈

Chao P'u-ch'ing 趙普卿

Chao Ya-tseng 趙亞曾

Ch'en Chih-mai 陳之邁

chen-li 真理

Ch'en I 陳儀

Ch'en T'ao-i 陳陶遺

Ch'en Tu-hsiu 陳獨秀

Cheng Chin 鄭錦

Cheng-ling 正豐

Chiang Meng-lin 蔣夢麟

Chiang Pai-li (Chiang Fang-chen) 蔣百里 (蔣方震)

Chiang-yüan k'ao 江源考

chien-she 建設

chih 質

chin shih 進士

Ching-hsing 井陘

ching-shen 精神

Ch'ing-k'uang-shan 青鑛山

Chou Mei-sheng 周枚生

Chou Sung-sheng 周頌聲

Chu Ching-nung 朱經農

Chu T'ing-hu 朱庭祜

Chu Yao-sheng 竹堯生

Chuang Wen-ya 莊文亞

chüeh-kuan kan-ch'u 覺官感觸

Chung-hsing 中興

Chung-kuo k'uang-yeh chi-yao 中國鑛業紀要

Chung-kuo ku-sheng-wu chih 中國古生物誌

Chung-yang tsu-jan li-shih po-wu-yüan 中央自然歷史博物院

Chung-yang yen-chiu yüan 中央研究院

Fan Shou-k'ang 范壽康
Fu Ssu-nien 傅斯年
Hsia Ch'ao 夏超
Hsiao chiang 小江
Hsiang tao 嚮導
Hsieh Chia-yung 謝家榮
Hsin ch'ao 新潮
Hsin ch'ing-nien 新青年
hsin-hsüeh 新學
hsin-shih 新式
hsin wen-hua 新文化
hsiu-ts'ai 秀才
Hsiung Hsi-ling 熊希齡
Hsü Hsia-k'o (Hsü Hung-tsu) 徐
　霞客 (徐宏祖)
Hsü Hsin-liu 徐新六
Hsü Ku-ch'ing 徐固卿
Hsü Pao-huang 徐寶璜
Hsü Yüan-hui 徐沅會
Hu Shih 胡適
Hua Heng-fang 華蘅芳
Hung-shui 洪水
i-ko 一個
Jen Shu-jung 任叔永
jen-sheng kuan 人生觀
Kai-tsao 改造
K'ang Yu-wei 康有為
Kao I-han 高一涵
k'ao-chü 考據
Ku Yen-wu 顧炎武
kuan tu shang pan 官督商辦
K'uang-cheng ssu 鑛政司
Ku-niu-chai 古牛寨
Kuo Mo-jo 郭沫若
Kuo-wen chou-pao 國聞週報
Kung shang pu 工商部
kung-sheng 貢生
Li Chi 李濟
Li Chi 李季
Li Chieh 李捷

Li Chien-tung 李建勳
li-hsüeh 理學
Li Hsüeh-ch'ing 李學清
Li Pao-chang 李寶章
Li Ssu-kuang 李四光
Li Ta-chao 李大釗
Li Tsu-hung 李祖鴻
Li Yüan-hung 黎元洪
Liang Ch'i-ch'ao 梁啓超
liang-chih 良知
liang-hsin 良心
Lin Tsai-p'ing 林宰平
Liang Sou-ming 梁漱溟
Liu Hou-sheng 劉厚生
Liu Tzu-k'ai 劉子楷
Lo Wen-kan 羅文幹
Lu Chih 陸贄
Lu-nan-shan 魯南山
Lung-chao-shan 龍爪山
Lung-yen 龍烟
Lung Yen-hsien 龍研仙
Man-yu san-chi 漫游散記
Mao Pang-wei 毛邦偉
Mei-chou p'ing-lun 每週評論
Mei Yüeh-han 梅月涵
mi-shu 秘書
Mien fang-chih-jan shih-yen-kuan
　棉紡織染實驗館
Min-kuo chün-shih chin-chi 民國軍事
　近紀
Min-tsu tsu 民族組
Nan-yang chung-hsüeh
　南洋中學
Nu-li chou-pao 努力週報
Nung shang pu 農商部
P'an-chiang k'ao 盤江考
Pei-p'iao 北票
P'eng Yün-i 彭允彝
P'ing-lun hui 評論會
P'u-tu ho 普渡河

Shao P'iao-p'ing 邵飄萍
Shang kung pu 商工部
She-hui k'o-hsüeh yen-chiu-so 社會科學研究所
She-hui t'iao-ch'a-so 社會調查所
Shih-ching-shan 石景山
Shih Chiu-yüan 史九元
shih-shih 事實
Sun Ch'uan-fang 孫傳芳
Sung-Hu shang-pu tsung-pan 淞滬商埠總辦
Sung Ying-hsing 宋應星
Tahing 泰興
Tai Chen 戴震
T'an Hsi-ch'ou 譚錫疇
T'an Ssu-t'ung 譚嗣同
T'ang Erh-ho 湯爾和
T'ao Chih-hsing 陶知行
T'ao Hsi-sheng 陶希聖
T'ao Hsiang 陶湘
T'ao Meng-ho 陶孟和
Ti-chih chuan-pao 地質專報
Ti-chih hui-pao 地質彙報
Ti-chih t'iao-ch'a-so 地質調查所
Ti-chih yen-chiu-so 地質研究所
t'i-yung 體用
T'ien kung k'ai wu 天工開物
Ting Ta-ko 丁大哥
Ting Wen-chiang (Ting Tsai-chün) 丁文江（丁在君）
Ting Wen-t'ao 丁文濤
Ting Wen-yüan 丁文淵
Ts'ai Yüan-p'ei 蔡元培
Tsiang T'ing-fu (Chiang T'ing-fu, T. F. Tsiang) 蔣廷黻
Ts'ao K'un 曹錕
Tseng Kuo-fan 曾國藩
Tseng Shih-ying 曾世英
Ts'ui Shu 崔述
tsung-kan-shih 總幹事

Tu-li p'ing-lun 獨立評論
Tuan Ch'i-jui 段祺瑞
Tung Ch'ang 董常
Tung-chuan 東川
t'ung-chih ching-chi 統制經濟
Tung Hsien-kuang (Hollington Tong) 董顯光
Tung K'ang 董康
T'ung-wen-kuan 同文館
Tzu-yu i-chih 自甲意志
Wang Chia-chü 王家駒
Wang Cheng 王徵
Wang Ching-ch'un 王景春
Wang Chu-ch'üan 王竹泉
Wang Ch'ung-hui 王寵惠
Wang Po-ch'iu 王伯秋
wen-jo 文弱
Wen-yu-hui 文友會
Wong Wen-hao (W. H. Wong) 翁文灝
Wu Tsung-chih 吳宗植
Wu-chang yü-ti hsüeh-hui 武昌輿地學會
wu-chih 物質
Wu Chih-hui 吳稚暉
Wu Ching-ch'ao 吳景超
Wu P'ei-fu 吳佩孚
Yang-ch'üan 陽泉
Yang Chü-ch'eng 楊聚誠
Yang Wei-hsin 楊維新
Yeh Liang-fu 葉良輔
Yen Fu 嚴復
Yen Hsi-shan 閻錫山
Yen Jo-chü 閻若璩
Yen Jen-kuang 顏任光
yu-hsiu fen-tzu 優秀分子
yu pi-jan-hsing 有必然性
Yü T'ung-k'uei 俞同奎
Yüan Shih-k'ai 袁世凱
yün ming ch'ien-ting 運命前定

293

Index

295

Harvard East Asian Series

20. *China's Wartime Finance and Inflation, 1937–1945.* By Arthur N. Young.
21. *Foreign Investment and Economic Development in China, 1840–1937.* By Chi-ming Hou.
22. *After Imperialism: The Search for a New Order in the Far East, 1921–1931.* By Akira Iriye.
23. *Foundations of Constitutional Government in Modern Japan, 1868–1900.* By George Akita.
24. *Political Thought in Early Meiji Japan, 1868–1889.* By Joseph Pittau, S.J.
25. *China's Struggle for Naval Development, 1839–1895.* By John L. Rawlinson.
26. *The Practice of Buddhism in China, 1900–1950.* By Holmes Welch.
27. *Li Ta-chao and the Origins of Chinese Marxism.* By Maurice Meisner.
28. *Pa Chin and His Writings: Chinese Youth Between the Two Revolutions.* By Olga Lang.
29. *Literary Dissent in Communist China.* By Merle Goldman.
30. *Politics in the Tokugawa Bakufu, 1600–1843.* By Conrad Totman.
31. *Hara Kei in the Politics of Compromise, 1905–1915.* By Tetsuo Najita.
32. *The Chinese World Order: Traditional China's Foreign Relations.* Edited by John K. Fairbank.
33. *The Buddhist Revival in China.* By Holmes Welch.
34. *Traditional Medicine in Modern China: Science, Nationalism, and the Tensions of Cultural Change.* By Ralph C. Croizier.
35. *Party Rivalry and Political Change in Taishō Japan.* By Peter Duus.
36. *The Rhetoric of Empire: American China Policy, 1895–1901.* By Marilyn B. Young.
37. *Radical Nationalist in Japan: Kita Ikki, 1883–1937.* By George M. Wilson.
38. *While China Faced West: American Reformers in Nationalist China, 1928–1937.* By James C. Thomson Jr.
39. *The Failure of Freedom: A Portrait of Modern Japanese Intellectuals.* By Tatsuo Arima.
40. *Asian Ideas of East and West: Tagore and His Critics in Japan, China, and India.* By Stephen N. Hay.
41. *Canton under Communism: Programs and Politics in a Provincial Capital, 1949–1968.* By Ezra F. Vogel.
42. *Ting Wen-chiang: Science and China's New Culture.* By Charlotte Furth.